TS'
227
N375
2005

New developments in advanced welding

Related titles:

The welding of aluminium and its alloys
(ISBN-13: 978-1-85573-567-5; ISBN-10: 1-85573-567-9)
A practical user's guide to all aspects of welding aluminium and aluminium alloys, giving a basic understanding of the metallurgical principles involved. The book is intended to provide for engineers, with perhaps little prior acquaintance with the welding processes involved, a concise and effective reference to the subject. It covers weldability of aluminium alloys; process descriptions, advantages, limitations, proposed weld parameters, health and safety issues; preparation for welding, quality assurance and quality control issues.

MIG welding guide
(ISBN-13: 978-1-85573-947-5; ISBN-10: 1-85573-947-X)
Gas metal arc welding (GMAW) also referred to as MIG (metal inert gas) is one of the key processes in industrial manufacturing. The *MIG welding guide* provides comprehensive, easy-to-understand coverage of this widely used process. The reader is presented with a variety of topics from the choice of shielding gases, to filler materials and welding equipment, and lots of practical advice. The book provides an overview of new developments in various processes such as: flux cored arc welding, new high productive methods, pulsed MIG welding, MIG-brazing, robotic welding applications and occupational health and safety.

Processes and mechanisms of welding residual stress and distortion
(ISBN-13: 978-1-85573-771-6; ISBN-10: 1-85573-771-X)
Through the collaboration of experts, this book provides a comprehensive treatment of the subject. It develops sufficient theoretical treatments on heat transfer, solid mechanics and materials behaviour that are essential for understanding and determining welding residual stress and distortion. It outlines the approach for computational analyses that engineers with sufficient background can follow and apply. The book will be useful for advanced analysis of the subject and provides examples and practical solutions for welding engineers.

Details of these and other Woodhead Publishing materials books and journals, as well as materials books from Maney Publishing, can be obtained by:

- visiting our web site at www.woodheadpublishing.com
- contacting Customer Services (e-mail: sales@woodhead-publishing.com; fax: +44 (0) 1223 893694; tel.: +44 (0) 1223 891358 ext.30; address: Woodhead Publishing Ltd, Abington Hall, Abington, Cambridge CB1 6AH, England)

If you would like to receive information on forthcoming titles, please send your address details to: Francis Dodds (address, tel. and fax as above; email: francisd@woodhead-publishing.com). Please confirm which subject areas you are interested in.

Maney currently publishes 16 peer-reviewed materials science and engineering journals. For further information visit www.maney.co.uk/journals.

New developments in advanced welding

Edited by
Nasir Ahmed

Woodhead Publishing and Maney Publishing
on behalf of
The Institute of Materials, Minerals & Mining

CRC Press
Boca Raton Boston New York Washington, DC

WOODHEAD PUBLISHING LIMITED
Cambridge England

Woodhead Publishing Limited and Maney Publishing Limited on behalf of
The Institute of Materials, Minerals & Mining

Published by Woodhead Publishing Limited, Abington Hall, Abington,
Cambridge CB1 6AH, England
www.woodheadpublishing.com

Published in North America by CRC Press LLC, 6000 Broken Sound Parkway, NW,
Suite 300, Boca Raton, FL 33487, USA

First published 2005, Woodhead Publishing Limited
© Woodhead Publishing Ltd, 2005
The authors have asserted their moral rights.

This book contains information obtained from authentic and highly regarded sources. Reprinted material is quoted with permission, and sources are indicated. Reasonable efforts have been made to publish reliable data and information, but the authors and the publishers cannot assume responsibility for the validity of all materials. Neither the authors nor the publishers, nor anyone else associated with this publication, shall be liable for any loss, damage or liability directly or indirectly caused or alleged to be caused by this book.

Neither this book nor any part may be reproduced or transmitted in any form or by any means, electronic or mechanical, including photocopying, microfilming and recording, or by any information storage or retrieval system, without permission in writing from Woodhead Publishing Limited.

The consent of Woodhead Publishing Limited does not extend to copying for general distribution, for promotion, for creating new works, or for resale. Specific permission must be obtained in writing from Woodhead Publishing Limited for such copying.

Trademark notice: Product or corporate names may be trademarks or registered trademarks, and are used only for identification and explanation, without intent to infringe.

British Library Cataloguing in Publication Data
A catalogue record for this book is available from the British Library.

Woodhead Publishing ISBN-13: 978-1-85573-970-3 (book)
Woodhead Publishing ISBN-10: 1-85573-970-4 (book)
Woodhead Publishing ISBN-13: 978-1-84569-089-2 (e-book)
Woodhead Publishing ISBN-10: 1-84569-089-3 (e-book)
CRC Press ISBN-10: 0-8493-3469-1
CRC Press order number: WP 3469

The publishers' policy is to use permanent paper from mills that operate a sustainable forestry policy, and which has been manufactured from pulp which is processed using acid-free and elementary chlorine-free practices. Furthermore, the publishers ensure that the text paper and cover board used have met acceptable environmental accreditation standards.

Typeset by Replika Press Pvt Ltd, India
Printed by TJ International Ltd, Padstow, Cornwall, England

Contents

	Contributor contact details	ix
1	**Gas metal arc welding**	**1**
	Y. ADONYI, LeTourneau University, USA and J. NADZAM, Lincoln Electric Company, USA	
1.1	Introduction	1
1.2	Advances in GMAW technologies	1
1.3	GMAW process measurement and control	7
1.4	GMAW of particular metals	9
1.5	GMAW hybrid processes and other developments	14
1.6	Future trends	18
1.7	References	18
2	**Tubular cored wire welding**	**21**
	D. WIDGERY, ESAB Group (UK) Ltd, UK	
2.1	Introduction: process principles	21
2.2	Equipment	23
2.3	Benefits	24
2.4	Materials used in tubular cored wire welding	25
2.5	Optimising productivity	33
2.6	Process control and quality	34
2.7	Applications	34
2.8	Troubleshooting	35
2.9	Advantages and disadvantages	36
2.10	Sources of further information and advice	38
2.11	References	38
3	**Gas tungsten arc welding**	**40**
	B. L. JARVIS, CSIRO Manufacturing & Infrastructure Technology, Australia and M. TANAKA, Osaka University, Japan	
3.1	Introduction	40

3.2	Principles	41
3.3	The A-TIG process	52
3.4	The keyhole GTAW process	64
3.5	Future trends	76
3.6	References	77

4	Laser beam welding	81

V. MERCHANT, Consultant, Canada

4.1	Introduction: process principles	81
4.2	Energy efficiency	85
4.3	Laser parameters: their measurement and control	87
4.4	Weld quality assurance	96
4.5	Advantages of laser beam welding	98
4.6	Suitability of laser beam welding	99
4.7	Process selection	100
4.8	Current laser beam welding applications	100
4.9	Related processes	102
4.10	Safety in laser beam welding	103
4.11	Future trends	104
4.12	Sources of further information and advice	108
4.13	References	110

5	Nd:YAG laser welding	113

M. NAEEM, GSI Group, UK and M. BRANDT, Swinburne University of Technology, Australia

5.1	Introduction	113
5.2	Laser output characteristics	113
5.3	The Nd:YAG laser	118
5.4	The laser as a machining tool	121
5.5	Laser welding with Nd:YAG lasers	125
5.6	Nd:YAG laser welding tips: process development	129
5.7	Nd:YAG laser welding of different metals	132
5.8	Control of Nd:YAG laser welding	144
5.9	References	156

6	New developments in laser welding	158

S. KATAYAMA, Osaka University, Japan

6.1	Introduction	158
6.2	Strengths and limitations of current laser welding technologies	159
6.3	New areas of research in laser welding	170

6.4	Advances in laser welding processes	180
6.5	Applications of laser welding	188
6.6	Future trends	190
6.7	References	191

7 Electron beam welding 198

U. DILTHEY, RWTH-Aachen University, Germany

7.1	Introduction	198
7.2	Basics of the process	200
7.3	Electron beam welding machines	206
7.4	Micro-electron beam welding	210
7.5	Non-vacuum electron beam welding	214
7.6	Quality assurance	220
7.7	Applications	226
7.8	References	227

8 Developments in explosion welding technology 229

J. BANKER, Dynamic Materials Corporation, USA

8.1	Introduction	229
8.2	Capabilities and limitations	229
8.3	EXW history	231
8.4	The EXW process	231
8.5	EXW applications	233
8.6	Weld characterization	238
8.7	Conclusions	239
8.8	References	240

9 Ultrasonic metal welding 241

K. GRAFF, Edison Welding Institute, USA

9.1	Introduction	241
9.2	Principles of ultrasonic metal welding	242
9.3	Ultrasonic welding equipment	252
9.4	Mechanics and metallurgy of the ultrasonic weld	254
9.5	Applications of ultrasonic welding	260
9.6	Summary of process advantages and disadvantages	262
9.7	Future trends	266
9.8	Sources of further information and advice	268
9.9	References	269

10	Occupational health and safety	270
	F. J. BLUNT, University of Cambridge, UK	
10.1	Introduction	270
10.2	Legislation	271
10.3	Recent and ongoing research	277
10.4	Environmental issues	282
10.5	Sources of further information and advice	286
10.6	References	288

Index 293

Contributor contact details

(* indicates main point of contact)

Editor

Dr Nasir Ahmed
CSIRO Manufacturing &
Infrastructure Technology
32 Audley Street
Woodville North
SA 5012
Australia

email: Nasir.Ahmed@csiro.au

Chapter 1

Professor Yoni Adonyi*
LeTourneau University
PO Box 7001
Longview, TX 75607
USA

email: yoniadonyi@letu.edu

Jeff Nadzam
Lincoln Electric Company
Cleveland, OH
USA

Chapter 2

Dr David Widgery
ESAB Group (UK) Ltd
Hanover House
Queensgate
Britannia Road
Waltham Cross
Herts EN8 7TF
UK

email: david.widgery@esab.co.uk

Chapter 3

Dr Laurie Jarvis*
CSIRO Manufacturing &
Infrastructure Technology
32 Audley Street
Woodville North
SA 5012
Australia

email: laurie.jarvis@csiro.au

Dr Manabu Tanaka
Joining and Welding Research Institute
Osaka University
11-1 Mihogaoka Ibaraki
Osaka 567-0047
Japan

email: tanaka@jwri.osaka-u.ac.jp

Chapter 4

Dr Vivian Merchant
13112 WestKal Road
Vernon BC
VIB IY5
Canada

email: thelaserguru@hotmail.com

Chapter 5

Professor Milan Brandt*
IRIS
Swinburne University of Technology
PO Box 218
Hawthorn
Victoria 3122
Australia

email: mbrandt@swin.edu.au

Dr Mohammed Naeem
email: naeemm@gsigrp.com

Chapter 6

Professor Seiji Katayama
Joining and Welding Research Institute
Osaka University
11-1 Mihogaoka
Ibaraki
Osaka 567-0047
Japan

email: katayama@jwri.osaka-u.ac.jp

Chapter 7

Professor Dr Ulrich Dilthey
ISF – Welding and Joining Institute
RWTH-Aachen University
Ponstrasse 49
D-52062 Aachen
Germany

email: di@isf.rwth-aachen.de

Chapter 8

Mr John G. Banker
Vice-President
Clad Metal Division
Dynamic Materials Corporation
5405 Spine Road
Boulder, CO 80301
USA

email: jbanker@dynamicmaterials.com

Chapter 9

Dr Karl Graff
Edison Welding Institute
1250 Arthur E. Adams Drive
Columbus, OH 43221-3585
USA

email: Karl_Graff@ewi.org

Chapter 10

Dr Jane Blunt
Department of Physics
Cavendish Laboratory
University of Cambridge
Madingley Road
Cambridge CB3 0HE
UK

email: fjb27@phy.cam.ac.uk

1
Gas metal arc welding

Y. A D O N Y I, LeTourneau University, USA and
J. N A D Z A M, Lincoln Electric Company, USA

1.1 Introduction

This chapter on gas metal arc welding (GMAW) assumes that the reader is already familiar with the fundamentals of the process. The review is divided into four sections based on process inputs, outputs, control systems and diverse advances in the GMAW process. Section 1.2 describes advances in power sources, wire electrode types, wire feeding and shielding gases. Section 1.3 includes a review of process analysis, sensing/monitoring, control, modelling, automation and robotics, simulations and arc physics/droplet transfer modes. Section 1.4 deals with process outputs such as microstructure/property relationships in ferrous and non-ferrous alloy welding. Section 1.5 reviews miscellaneous GMAW-related improved processes such as hybrid laser/GMAW, tandem GMAW welding, narrow groove GMAW welding and digital networking of power sources. Finally, future trends are predicted based on recent advancements in the GMAW process simulation, modelling, sensing and control.

An earnest effort has been made to incorporate recent information available in the open technical literature. The authors would like to apologise to those technical experts whose work might have been omitted by mistake, oversight or lack of availability of published papers. Special thanks go to Dr Karin Himmelbauer from Fronius International, Wels, Austria and Dr Prakriti Kumar Ghosh from the Indian Institute of Technology, Roorkee, India, for their significant contributions to this chapter.

1.2 Advances in GMAW technologies

1.2.1 Power sources

Traditional power sources for GMAW welding are the analogue constant voltage (CV) type, with the welding current setting controlled by the wire electrode feeding rate (Nadzam, 2003). A schematic of a typical DC power source, with the welding torch connected to the positive electrode or DC+ or DCRP is shown in Fig. 1.1. As a reminder, the fundamentals of the GMAW process are presented schematically in Fig. 1.2, showing the electric arc, gas shielding, wire electrode and weld deposit.

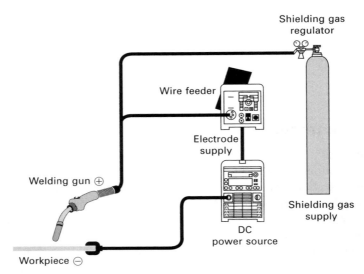

1.1 Schematic view of a GMAW welding system showing main components (Himmelbauer, 2003).

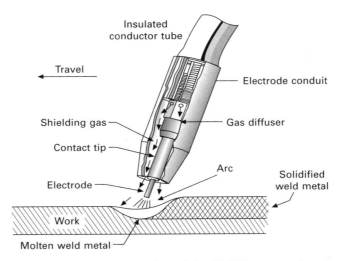

1.2 Schematic representation of the GMAW process in a longitudinal cross-section (Nadzam, 2003).

Gas metal arc welding

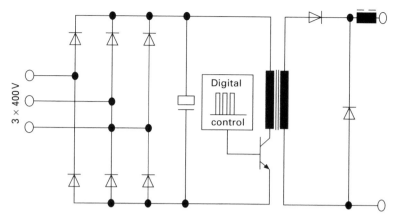

1.3 Schematic representation of electrical components of a typical GMAW digital power source and control system (Himmelbauer, 2003).

One major development in improving the efficiency of the transformer introduced in the 1980s was in using high frequencies to reduce thermal energy losses via eddy current heating of the transformer core, thus reducing the size of the transformers. The current in the secondary was then lowered again for welding and this 'inverter' technology was used to make GMAW power sources more portable (Fig. 1.3, Himmelbauer, 2003). Furthermore, as digital control technology improved, pulsed GMAW (GMAW-P) power sources were developed, with a block representation shown in Fig. 1.4. With these digital power sources several improvements were accomplished besides better process control and reproducibility: the ability to programme and monitor the waveform, remote access and single-knob ('synergic') adjustment with the control panel shown in Fig. 1.5 (Courtesy Fronius International). Using this control, sets of pre-programmed welding parameters are called out from a large database, eliminating the trial and error set-up typical of semi-automatic operations.

1.2.2 Wire feeding

Different wire electrode types can have specific problems with feeding in 'push' and 'push–pull' modes. Mathematical modelling and experiments recently showed that the friction force between the wire and its liner resisting feeding increases exponentially with the liner bend angle (Padilla *et al.*, 2003). To reduce this friction force, a typical push–pull torch is shown in Fig. 1.6, while another solution to this feeding problem is to use small spools attached to the torch (Fig. 1.7), (Nadzam, 2003).

Contact tube life can be extended by understanding better the thermal deterioration process governing its wear. It was found that the radiant heat of

4 New developments in advanced welding

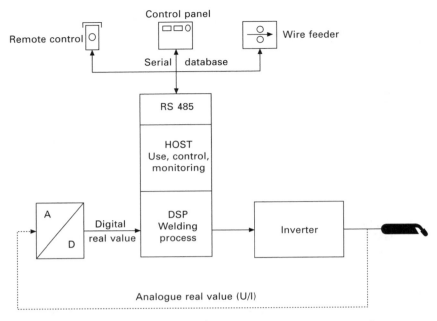

1.4 Block diagram of a modern inverter-type power supply (Himmelbauer, 2003).

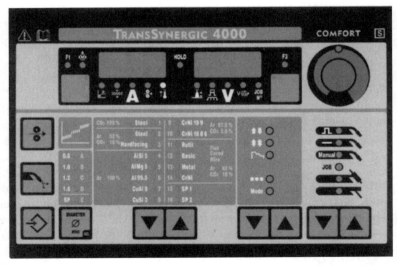

1.5 View of a 'single-knob' control panel for a typical synergic GMAW system (Himmelbauer, 2003) (Courtesy Fronius International).

Gas metal arc welding 5

1.6 View of a typical 'push–pull' GMAW welding gun that incorporates an extra wire feeding motor in the handle.

1.7 GMAW welding pull-type gun having a small diameter wire spool attached to the gun (Nadzam, 2003).

the arc plasma and resistive heating at the electrode–wire interface are mostly responsible for heating the contact tip (Adam *et al.*, 2001). Thus the contact tip to work distance CTWD (Fig. 1.8) had the most important effect on the tip overheating, while arc-on time also played a major role on contact tip temperature. The lower the CTWD, the more overheating the electrode tip experienced, confirming the major role of heat radiation from the arc on the contact tip temperature and consequent wear.

1.2.3 Wire electrode geometry

Traditionally, solid cylindrical wires have increasingly been replaced by tubular electrodes, i.e. metal-core or flux-core (Myers, 2001). The main advantage of these cored wires lies both in their containing a mix of alloying

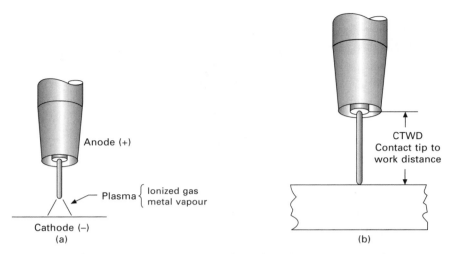

1.8 Schematic representation of the GMAW process (a) and control dimension CTWD (b).

elements and in their flexibility for tailoring weld deposit properties. Metal core (MC) wires have been used to weld high performance weathering steels with 70 and 100 ksi (490 and 700 MPa) yield strengths and excellent toughness at no preheating in 50.8 mm (2 inch) thick plates. Use of large diameter solid wires of up to 3.2 mm (0.125 inch) in diameter resulted in increased deposition rates at equal power (Himmelbauer, 2003). Another development is in use of strip wires of 0.5 × 4.5 mm rectangular cross-section instead of cylindrical ones (Himmelbauer, 2003). One major advantage of using these strip wires is in higher wire feed speeds up to 11 m/min and therefore the high deposition rate by using a push–pull system (Fig. 1.9). Penetration was lower when compared with round cross-sections of equivalent area, but strip wires can therefore easily be used for weld surfacing. One major disadvantage of strip wires lies in attempting to feed them in twisted liners typical of robotic and complex semi-automatic motions.

1.2.4 Shielding gases

Two major types of shielding gases are being used in GMA welding: (1) inert and (2) active or reactive. European standards designate the two subsections of GMAW as MIG (metal inert gas) vs. MAG (metal active gas) welding. Binary and ternary gas mixes have been developed in order to optimise the chemical activity, ionization potential and thermal conductivity combination (Vaidya, 2001; Zavodny, 2001). Application of these custom-made gas mixes also have to be co-ordinated with the droplet transfer modes used (Nadzam, 2003). Care has to be exercised when using Ar + CO_2 mixtures in welding

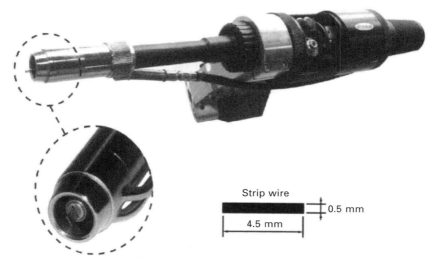

1.9 Rectangular strip wires used with 'push–pull' feeding systems (Himmelbauer, 2003).

stainless steels, as detrimental carbon pickup can occur (Kotecki, 2001). There can be an adverse effect of metal transfer mode on the weld carbon contamination; the worst is the spray mode for a given CO_2 content in the shielding gas.

1.3 GMAW process measurement and control

This section includes a review of recent advances in GMAW process analysis. Topics include process sensing/monitoring, control, modelling, automation and robotics, simulations, arc physics/droplet transfer modes and fume and spatter control.

1.3.1 Droplet transfer modes

One of the major topics in GMAW process analysis has been the molten metal droplet detachment and transfer modes. For given ranges of wire electrode diameter, welding current and shielding gas, five modes of detachment have been recognised (Nadzam, 2003): (1) short-circuit, (2) globular, (3) axial spray, (4) pulsed-spray and (5) surface-tension transfer modes (Fig. 1.10) (Nadzam, 2003).

The forces governing the dynamic equilibrium during droplet detachment have been identified. They are: (a) electromagnetic forces associated with the welding current self-induced magnetic field, (b) gravity, (c) surface tension and (d) cathodic jet forces (Lancaster, 1984). Lately, variable polarity

8   New developments in advanced welding

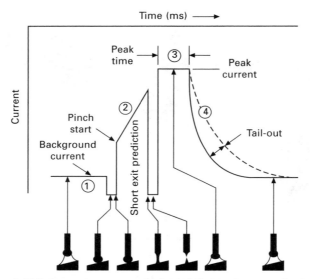

1.10 Schematic representation of the surface tension-controlled droplet transfer mode (STT) (Nadzam, 2003).

(VP-GMAW) has been shown to be effective also in controlling metal transfer and melting rate.

1.3.2 Process control

Traditionally, using a CV power source with inductance control proved to be excessively sensitive to arc length variations, responding with large wire speed and current responses. Therefore, feed-forward controls – also known as digital or reactive controls – have been introduced where the current can be modified independently from the wire speed. Advances in process control have been made especially using feed-forward algorithms, as demonstrated by their excellent adaptability to step responses when compared to the traditional feed-back control (Adolfsson, 1999). A resulting constant arc length control system is schematically shown in Fig. 1.11 (Himmlebauer, 2003), demonstrating the ability of the system to adjust to random variations in CTWD without changing the arc length.

Process control can also be very different in aluminium alloys when compared to that in steel. For the same wire electrode extension, the Al GMAW was found to be up to 28 times more sensitive to variations in wire feed speed than the mild steel electrode (Quinn, 2002). Because of the higher electrical and thermal conductivity of Al compared to steel, conductive heat transfer dominates the dynamic equilibrium between burn-off and feed rate, compared to convection and resistive heating in steels. For instance, the

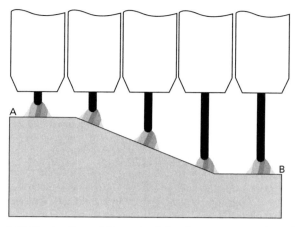

1.11 Illustration of the principle of constant arc length control when using digital power systems (Himmelbauer, 2003).

voltage drop across the same electrode extension length was one order of magnitude less in Al than it was steels (0.03 V compared to 0.3 V). In most cases, the GMAW process responds in aluminum more dramatically to perturbations in welding current or wire speed setpoints.

Process stability in pulsed GMAW of titanium can be improved by active droplet transfer control by adding peak current pulse in the one-drop-per-pulse or ODPP method or excited droplet oscillation (Zhang and Li, 2001). Application of statistical process design has been used in the past decade to optimise GMAW welding parameter development (Allen *et al.* 2002). Methods involve classical design-of-experiments or DOE, heuristic parameter optimisation, neural network modelling and Taguchi methods. Generally, independent and dependent variables are identified using regression analysis of the weld quality and empirical equations are developed to predict optimum parameters for new situations. Invariably, the original experiments are limited to the base material type, joint design and fitup, shielding gas type, etc. and most such articles end up apologising for the narrow range of applicability of their predictions (Subramaniam *et al.*, 1999).

1.4　GMAW of particular metals

This section attempts to describe GMAW process outputs such as microstructure/property relationships in ferrous and non-ferrous alloy welding.

1.4.1　Microstructure/property relationships

Although pulsed current power sources (GMAW-P) have originally been developed to improve process stability, penetration and deposition rates, it

10　New developments in advanced welding

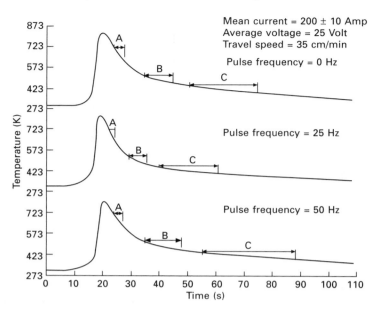

1.12 Increased pulsed frequency reducing cooling rate of HAZ adjacent to fusion line of Al-alloy weld (Ghosh *et al.*, 1994).

was also found that they could control the weld deposit properties. The solidification mechanism of weld metal during GMAW-P welding differs from that of continuous current welding due to intermittent movement of heat towards the solidification front. During pulsed current GMA welding, the solidification of the molten pool takes place primarily in two steps: one during the pulse off time period, and the second during development of a weld spot resulting from the next pulse. Thus the weld deposit microstructures and heat affected zone (HAZ) width can be varied (Ghosh *et al.*, 1990a) as GMAW-P waveform types can produce a wide range of metal transfer energy levels, deposition rates and resulting thermal cycles (Gupta *et al.*, 1988) (Fig. 1.12). By this course of action the pulsed current GMA welding improves the weld property in comparison with that of the conventional continuous current GMA weld (Ghosh *et al.*, 1990b) (Fig. 1.13).

In out-of-position GMAW-P, the appropriate selection of pulse parameters, such as mean current (I_m), peak current (I_p), base current (I_b), pulse duration (t_p) and pulse frequency (f), in combination provides a droplet velocity. The droplet is propelled or rejected by gravity, depending on the welding position, and it imposes a control over the superheated droplet and the resulting fluidity of the weld pool. Successful use of pulsed current GMA welding to produce weld of desired quality is very much dependent upon proper control over the I_m I_p, I_b, f, t_p and pulse off (base) time (t_b). Because of the interrelated nature

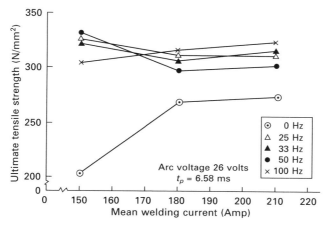

1.13 Effect of pulsing on the tensile strength in GMAW-P when compared to continuous wave (0 Hz). (Ghosh *et al.*, 1991).

of pulse parameters, control of weld quality is possible by establishing a correlation with a dimensionless factor $\phi = [(I_b/I_p)ft_b]$, where t_b is the pulse-off time, expressed as $[(1/f)-t_p]$.

1.4.2 GMAW-P welding of C–Mn steels

It was found that in case of FCAW-P bead-on-plate deposits, the pulse parameters and changes in arc voltage affects the microstructure and hardness of the weld deposit and HAZ, the width of HAZ and increases the porosity content of the deposit (Ghosh and Rai, 1996) (Fig. 1.14). The pulse parameters and arc voltage (due to their influence on I_p and I_b) are found to affect the characteristics of the weld deposit and HAZ via the factor ϕ. At a given welding speed and arc voltage the width and hardness of HAZ show a relatively decreasing and increasing trend in the linear relationship with the factor ϕ. Optimised pulse parameters can improve the weld quality, i.e. reduce porosity and optimise microstructure/hardness in the weld deposit and HAZ. GMAW-P in the welding of large diameter cross-country pipelines is useful for reducing the occurrence of incomplete fusion defects. The superiority of using the GMAW-P process over the short-circuiting arc conventional GMAW process in a vertical-up weld deposition has been marked by a significant enhancement of the tensile, impact and fatigue properties of the weld joint of C–Mn structural steel. The variation in microstructures of the weld metal and HAZ, the geometry of the weld deposit, and the properties of the joint with a change in pulse parameters maintains a good correlation to the factor ϕ. It has been reported that, at a given arc energy, the variation in ϕ significantly influences the morphology of the weld deposit, becoming finer with increased

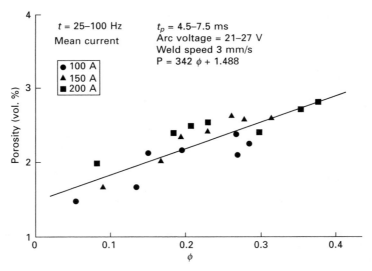

1.14 Effect of pulse parameter on porosity content of pulsed current GMA weld deposit of C–Mn steel (Hussain *et al.*, 1999).

ϕ. Showing a general tendency to fracture from the weld deposit, the fatigue life of a pulsed-current weld becomes higher than that of the conventional short-circuiting arc GMA weld. Again, the most likely reason for this is the greater hardness of the weld deposit and the corresponding tensile strength. As fatigue failure initiation life has been known to be proportional to the tensile strength, it is probable that an increase in ϕ had an indirect effect on fatigue life. Reduced net heat input in GMAW-P resulted in increased cooling rates and higher HAZ hardness in high performance steels, when compared to continuous wave GMAW welding using the same calculated arc energy (Adonyi, 2002).

1.4.3 GMAW-P welding of Al alloys

Variation of pulse parameters in single and multi-pass pulsed current GMA welding of Al alloys up to 25 mm thickness significantly affected the geometry, microstructure, and mechanical properties of weld joints (Ghosh *et al.*, 1990a; Ghosh *et al.*, 1990b; Ghosh and Hussain, 2002). GMAW-P at an average current level above the globular-to-spray mode transition current of the filler wire with suitable combination of pulse parameters significantly refines the microstructure of weld deposit and reduces the width of recrystallized HAZ (Ghosh *et al.*, 1990a), when compared to continuous current GMAW deposited at the same average current level. Again, the most likely reason for this behaviour is the increase in HAZ cooling rates and lower peak temperatures attained within a given distance from the fusion line.

GMAW-P welding has also been found to improve the tensile strength and ductility of the weld joint of Al–Zn–Mg alloys in comparison to those of its conventional GMA weld, where the failure usually initiates in the Al–Mg weld deposit. The improvement in weld properties is primarily attributed to the refinement of the microstructure, reduction in porosity content and resulting weld geometry, which favourably control the dilution and zinc pickup from the base metal, forming strengthening precipitates in the weld deposit as identified by the X-ray diffraction studies. It is believed that GMAW-P improves the fatigue life of the weld by influencing the m and C values of the Paris law of crack growth rate expressed as $da/dN = C(\Delta K)^m$ where a is the crack length, N is the number of loading cycle, ΔK is the range of stress intensity factor and C and m are the material constants (Ghosh et al., 1994). The improvement in fracture mechanics properties is primarily attributed to the synergic effects of refinement of microstructure and amount of zinc pickup due to dilution of Al–Mg weld deposit by the base metal (Ghosh et al., 1991; Ghosh et al., 1994; Hussain et al., 1997; Hussain et al., 1999).

1.4.4 GMAW-P in stainless steel cladding

In single pass stainless steel cladding on structural steel plates the use of pulsed current GMAW-P, produces comparatively increased clad layer thickness, lowered penetration when compared to overlays produced by continuous current GMAW. Overlaying using GMAW-P also enhances the hardness of the clad layer and reduces the hardness of diffusion layer formed at the interface of stainless steel cladding with the structural steel (Ghosh

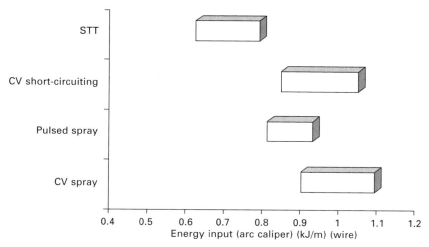

1.15 Net heat input for different molten metal droplet modes (Hussain et al., 1999).

et al., 1998). It can be assumed that all these effects are caused by the lower net heat input in GMAW-P and accelerated cooling rates, when compared to continuous GMAW with equivalent average arc energy. Lower net heat inputs of up to 20% (Hsu and Soltis, 2002) and 15% (Joseph *et al.*, 2002) were found using calorimetry and the arc instantaneous power measurements. Note that the initial arc efficiency was ~70% for the GMAW process in general (Fig. 1.15). Similar lower heat inputs relative to the arc energy were found in GMAW-P.

1.5 GMAW hybrid processes and other developments

This section reviews miscellaneous GMAW-related improved processes such as hybrid laser/GMAW, tandem GMAW welding, narrow groove welding and digital power source networking.

1.5.1 Hybrid welding processes: GMAW/LBW

This combination of high penetration laser beam welding (LBW) and good gap bridgeability (GMAW) processes builds on the intelligent combination of the advantages of each process. The resulting welds (Staufer *et al.*, 2003) can be made at high speeds, have good penetration and are less sensitive to gap variations. The GMAW arc stability and droplet transfer are also improved by the intense metal vaporisation caused by LBW. Apparently, the greater amount of ionised metal and electrons in the LBW plasma reduces the need for high ionisation potential and exceeding the electrode work function in the GMAW arc thus provides better arc stability. Disadvantages include: high capital cost and the need for automation and precise beam/arc alignment. Typical GMAW/LBW heads are expensive and complex, as shown in Fig. 1.16.

1.5.2 GMAW brazing

Using low-melting point electrode wire consumables such as Cu-Si, Cu-Ag, Cu-Al alloys allow for low current GMAW-P deposition without melting of the base metal (i.e. electric brazing). This significant development reduces the width of HAZs and damage to Zn coatings in the automotive sheet and produces minimal distortions (Himmelbauer, 2003). The roof panel joint does not require any post weld processing. Additionally, arc-brazing is also being accomplished using traditional and STT^{TM} forms of GMAW-S. GMAW with a CuAl wire electrode also makes possible joining of dissimilar materials with very different melting points such as steel and aluminium.

Gas metal arc welding   15

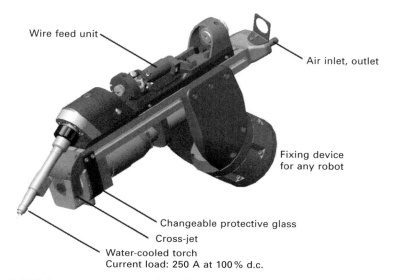

1.16 Schematic representation of a laser beam welding/GMAW hybrid welding head (Himmelbauer, 2003).

One disadvantage of GMAW brazing is the low joint strength that can be compensated by using lap joint design. Additional problems have been associated with zinc pickup in the silicon bronze weld and the result is transverse cracking of the weld deposit. This occurs in welds of those members where there is a gap. The gap, via capillary action picks up zinc from both surfaces of the plated base material. Finally, the presence of Cu in the recycled car bodies lowers the quality of the scrap and increases cost because of the difficulty of removing Cu which is very detrimental in steel making (solidification cracking susceptibility).

1.5.3 GMAW tandem welding

As the name implies, two wire electrodes are used in tandem to produce welds. The two wire electrodes are insulated from each other in tandem welding (Himmelbauer, 2003), thus the droplet transfer mode can be adjusted independently, in contrast to double-wire welding. Typically, one electrode can work in continuous arc (synergic CV or synergic CC) and the other in pulsed arc mode (also known as 'master' and 'slave' wires or 'lead' and 'trail' wire). Accordingly, the modified process allows for great flexibility in addition to increased travel speed, higher deposition rates, as well as lower spatter. Disadvantages include equipment complexity, as well as the need for automation. Seam tracking may or may not be required (Fig. 1.17). The system employs two power sources, two wire drives, and a control. It is

16   New developments in advanced welding

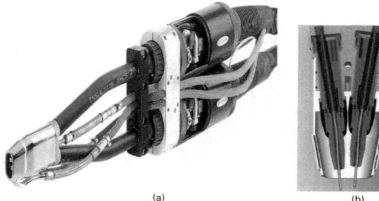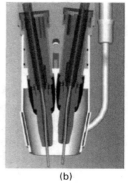

1.17 Tandem GMAW torch view (a) and cross-section (b).

adapted for either repetitive side-beam type applications or is employed with a welding robot. This variant of the gas metal arc welding process is capable of higher travel speeds, 1.5–2.0 times the speed of a single electrode. Some travel speeds may exceed 150 in/min (3.81 m/min). Deposition rates of 42 pounds/h (19.1 kg/h) are achievable for heavier plate welding (Nadzam, 2003).

The modes of metal transfer used for the tandem GMAW are axial spray metal transfer or pulsed spray metal transfer. The combinations of the modes that are popularly employed include:

- Spray + pulse: Axial spray transfer on the lead arc followed by pulsed spray transfer on the trail arc.
- Pulse + pulse: Pulsed spray transfer on both the lead and the trail arc.
- Spray + spray: Axial spray transfer on both the lead and the trail arc.

The higher energy spray + spray configuration is used for special heavy plate welding where deeper penetration is required. Pulse + pulse allows for heavy welding or high-speed sheet metal welding.

Central to the successful operation of tandem GMAW is proper understanding of the set-up of the special tandem GMAW welding torch. In most cases, the central axis of the torch should be normal to the weld joint. The lead arc has a built in 6 degree lagging electrode angle, and the trail has a built in 6 degree leading electrode angle.

The contact tip to work distance (CTWD) for higher speed sheet metal type applications should be set at 0.625 in (16 mm). The electrode spacing is critical and the shorter CTWD establishes the correct spacing. When the CTWD is held at this position the two arcs become more distinct from one another and shorter arc lengths are used to provide higher travel speeds. Use of tandem GMAW for heavy plate fabrication requires a longer CTWD,

1.0 in (25.4 mm). The longer CTWD provides the correct spacing between the two arcs, and in this scenario, the arcs tend to move very closely together. When held at the longer CTWD the arcs lend themselves for use with much higher wire feed speeds.

1.5.4 Narrow groove GMAW welding

An excellent application of GMAW is for low heat input welding of thick plate, the resulting welds have often been plagued by occasional lack of sidewall fusion. Wire electrode bending and rotating (twisted wire) have been used in the past to overcome this problem. Korean researchers used electromagnetic arc oscillation to alleviate the same problem in the narrow groove (Khang and Na, 2003).

1.5.5 Use of the World Wide Web and digital networks

Digital power sources and the associated data acquisition systems make possible linking them into local area networks (LAN) (Fig. 1.18). Depending on the availability of Internet connections at remote locations, qualified parameters can be directed to individual welding machines. On the other hand, welder procedure qualifications can be sent electronically to one central location.

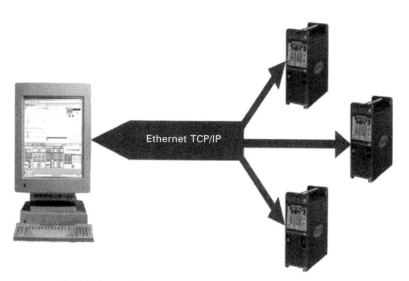

1.18 Schematic representation of data interchange between GMAW power sources and a CP using internet protocol via Ethernet (Himmelbauer, 2003).

1.6 Future trends

The following trends can be anticipated in GMA welding within the next five years. The following areas are important:

(1) process simulation and modelling,
(2) sensing and control,
(3) cost reduction and
(4) new applications.

- Improved computer *simulations* of the welding process and implementation in production welding;
- Improved *sensing* and signal acquisition before, during and after welding and inclusion in a comprehensive control system. This effort will require increased sensitivity to downstream manufacturing practices to improve part fitup;
- Improved power source technology via digital controls and improved control of the welding arcs;
- *Applications*: extension of the process to reduced base metal thickness and higher deposition rates. Even further miniaturisation (or MEMS: micro-electro-mechanical systems) can be expected to penetrate the GMAW equipment world;
- *Automation*: remote operation (depths, heights, hazardous environments);
- In semi-automatic applications: integration of all essential functions in the welding torch;
- Deposition rates and cost reductions – more hybrid and new process variants, lower cost filler wires and shielding gases (push toward self-shielded fluxed core arc welding;
- *Controls*: digital networks, qualifications.

1.7 References

Adam G., Siewert T.A., Quinn T.P. and Vigliotti D.P. (2001), 'Contact tube temperature during GMAW', *Welding Journal*, Dec. 2001, 37–41

Adolfsson S., Bahrami P. *et al.* (1999), 'On line quality monitoring in short circuit GMA welding', *Welding Journal, Research Supplement*, 59s–73s, Feb. 1999

Adonyi Y. (2002), 'Welding process effects in hydrogen induced cracking susceptibility of high performance steels', *Welding Journal, Research Supplement*, 61s–68s, Apr. 2002

Allen T.T. and Richardson R.W. *et al* (2002), 'Statistical process design for robotic GMA welding of sheet metal, *Welding Journal, Research Supplement*, 69s–77s, May 2002

Ghosh P.K. and Rai B.K. (1996), 'Characteristics of pulsed current bead on plate deposit in flux cored GMAW process', *ISIJ Int.*, **36**(8), 1036–45

Ghosh P.K., Gupta S.R., Gupta P.C. and Rathi R. (1990a), 'Influence of pulsed MIG welding on the microstructure and porosity content of Al–Zn–Mg alloy weldment', *Practical Metallography*, **27**, 613–26

Ghosh P.K., Gupta S.R., Gupta P.C. and Rathi R. (1990b), 'Pulsed MIG welding of Al–Zn–Mg alloy', *Materials Trans. JIM*, **31**(8), 723–9

Ghosh P.K., Gupta P.C., Gupta S.R. and Rathi R. (1991), 'Fatigue characteristics of pulsed MIG welded Al–Zn–Mg alloy', *J. Mater. Sci.*, **26**, 6161–70

Ghosh P.K., Dorn L. and Issler L. (1994), 'Fatigue crack growth behaviour of pulsed current MIG weld of Al–Zn–Mg alloy', *Int. J. Join. Mater.*, **6**(4), 163–8

Ghosh P.K., Gupta P.C. and Goyal V.K. (1998), 'Stainless steel cladding of structural steel plate using pulsed current GMAW process', *Weld. J., AWS*, **77**(7), 307–12s

Ghosh P.K. and Hussain H.M. (2002), 'Morphology and porosity content of multipass pulsed current GMA weld of Al–Zn–Mg alloy', *Int. J. Join. Mater.*, **41**(1/2), 16–26

Gupta P.C., Ghosh P.K. and Vissa S. (1988), 'Influence of pulse frequency on the properties of HAZ in pulsed MIG welded Al–Zn–Mg alloy', *Proc. Int. Conf. on Welding Technology in Developing Countries – Present Status and Future Needs*, September 26–28, (1988), pp. 71–77

Himmelbauer K., (2003), 'Digital Welding', Fronius International, Proprietary reports and presentations

Hsu C. and Soltis P. (2002), 'Heat input comparison of SST vs. short circuiting and pulsed GMAW vs. CV processes', *Sixth International Conference on Welding Research*, Pine Mountains, GA, 2002

Hussain H.M., Ghosh P.K., Gupta P.C. and Potluri N.B. (1997), 'Fatigue crack growth properties of pulse current multipass MIG weld of Al–Zn–Mg alloy', *Trans. Ind. Inst. Met.*, **50**(4), 275–85

Hussain H.M., Ghosh P.K., Gupta P.C. and Potluri N.B. (1999), 'Fracture toughness of pulse current multipass GMA weld of Al–Zn–Mg alloy', *Int. J. Join. Mater.*, **11**(3), 77–88

Joseph A., Harwig D.D., Farson D. and Richardson R. (2002), Assessing the effects of GMAW-P parameters on arc power and heat input', *EWI Report*

Khang Y.H. and Na S.J. (2003), 'Characteristics of welding and arc signal in narrow groove GMAW using electromagnetic arc oscillation', *Welding Journal, Research Supplement*, **82**(15), 93s–9s, May 2003

Kotecki D.J. (2001), 'Carbon pickup from argon-CO_2 blends in GMAW', *Welding Journal*, 43–8, Dec. 2001

Lancaster J.F. (1984), *The Physics of Welding*, London, International Institute of Welding and Pergamon Press

Myers D. (2001), 'Metal cored wires: advantages and disadvantages', *Welding Journal*, 39–42, Dec. 2001

Nadzam J., (2003), *Gas Metal Arc Welding: Process Overview*, Lincoln Electric Company, Technology Center, Internal report

Padilla T.M., Quinn T.P., Munoz D.R. and Rorrer R.A.L. (2003), Mathematical model of wire feeding mechanisms in GMAW welding', *Welding Journal, Research Supplement*, 100s–109s, May 2003

Quinn T.P. (2002), 'Process sensitivity in gas metal arc welding of aluminum vs. steel', *Welding Journal, Research Supplement*, 55s–60s, Apr. 2002

Staufer H., et al., (2003), *Laser Hybrid Welding and LaserBrazing: State-of-the-art in Technology and Practice: Audi A8 and VW paeton*, Internal Publication, Fronius International GmbH, Wels, Austria

Subramaniam D.R., White J.E. et al., (1999), 'Experimental approach to selection of pulsing parameters in pulsed GMAW', *Welding Journal, Research Supplement*, 166s–172s, May 1999

Vaidya V.V. (2001), 'Shielding gas mixtures for semiautomatic welds', *Welding Journal*, 43–8, Sep. 2002

Zavodny J. (2001), 'Welding with the right shielding gas', *Welding Journal*, 49–50, Dec. 2001

Zhang Y.L., Li P.J. (2001), 'Modified active control metal transfer and pulsed GMAW in titanium', *Welding Journal, Research Supplement*, 54s–61s, Febr. 2001

2
Tubular cored wire welding

D. WIDGERY, ESAB Group (UK) Ltd, UK

2.1 Introduction: process principles

Tubular cored wire welding was foreseen in the 1911 patent application[1] in which Oscar Kjellberg introduced the world to the concept of the coated welding electrode: as an alternative, he said, a tube could be used with the powder or paste inside. By 1936, 10 000 tonnes of tubular electrodes had been produced in Austria and the Schlachthof Bridge in Dresden had been fabricated by mechanised welding with tubular wire.[2] This advanced welding process thus has a rather longer history than some of the others described in this book. However, it was only in the last part of the twentieth century that competitive pressures spurred on the rapid expansion in its use that led to its important position today.

Tubular wires were developed to bring together the advantages of two existing processes. Manual metal-arc (MMA) electrodes have a coating which can alloy and deoxidise the weld, form a slag to protect, refine and support the pool, and contribute ionic species to stabilise and modify the arc. Gas-shielded metal-arc welding (GMAW) was introduced in the 1920s to allow continuous welding with its inherently greater productivity, but was limited, especially in positional welding, by its lack of slag. Tubular wires use in many cases the same welding equipment as solid wires but have a number of advantages in usability, productivity and metallurgical flexibility.

At this point it may be useful to clarify the terminology of tubular wires. Many of these are widely known as flux-cored wires, and the American Welding Society refers to 'flux-cored arc welding'. In the 1950s, a range of wires appeared containing no fluxing agents, but only metal powders: these became known as metal-cored wires. In American standards they are regarded as a subset of solid wires. However, British patents on metal-cored wires allowed for the inclusion of non-metallic elements in the core up to a total of 4% by weight,[3] and the European approach has been to see a continuum between metal-cored and flux-cored wires. Hence the term 'tubular wires' has been adopted in British and subsequently European and ISO standards.

New developments in advanced welding

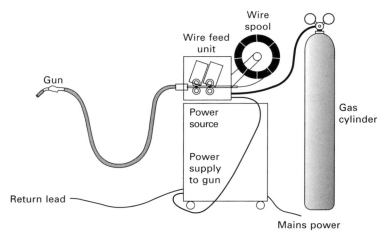

2.1 Equipment used for welding with tubular wire.

In operation, as seen in Fig. 2.1, a wire is fed from a spool through a conduit to a torch or gun, which may be hand-held or mounted on a mechanical traverse. The wire passes through a contact tip by means of which current is supplied to the wire, and which may be surrounded by a cylindrical gas nozzle or shroud if the wire is of a type that calls for gas shielding. As in gas-shielded welding with solid wire, the power source normally supplies an almost constant voltage so that the arc is self-stabilising. If, for example, the arc starts to shorten, its impedance is reduced. If the impedance of the power source is still lower, in other words if it is of the constant voltage type, coupling between the power source and the arc will increase and the wire will burn off faster until equilibrium is restored.

Users of gas-shielded welding with solid wire will be familiar with the different operating modes of the process as the wire feed speed and voltage vary. When both of these are high, fine droplets stream from the wire tip in what is known as 'spray transfer', so current and voltage remain relatively constant over both short and long time scales. At low voltages and currents, however, the arc power is not enough to burn off the wire as fast as it is feeding. The wire tip will approach the molten pool and eventually a short circuit will occur. Because of the sudden improvement in impedance matching, the power transfer increases and soon melts the wire tip. The molten bridge breaks, the arc is re-established and the process repeats itself. This is known as 'dip transfer'. Because the total power absorbed is less and the weld pool smaller, this transfer mode is preferred for positional welding.

In principle, similar transfer modes can be observed with tubular wires, albeit not always in such a pure form. Metal-cored wires can behave very like solid wires, with a transition from dip to spray. Basic flux-cored wires certainly have a short-circuiting transfer mode, but at higher currents the droplets rarely become fine enough for the transfer to be classified as true spray, even though it may appear as such to the welder. Rutile flux-cored wires, on the other hand, produce free-flight transfer over their whole operating range and do not rely on a short-circuiting mode for positional welding. These characteristics are dealt with in more detail in Section 2.4 below.

2.2 Equipment

Power sources were originally transformer–rectifier machines, later with thyristor control, but many today make use of inverters to allow the power transformer to operate at a higher frequency and thus be reduced in size. A further benefit is that the static and dynamic characteristics of the power source can be controlled more accurately and over a wider range using a microprocessor, so that the output can be optimised for each type of wire, and indeed often for covered electrodes as well. For tubular wires, operating voltages of 15–35 V may be needed. Unlike solid wires, some tubular wires are designed to operate with electrode negative and so may not be suitable for some older equipment which does not provide for this.

With a modern inverter set, it is often possible to provide a pulsed arc facility at little or no extra cost, and this can be helpful to reduce spatter when using basic flux-cored wires. Because the electrical and arc characteristics of tubular wires vary much more than those of solid wires, some experience may be needed to optimise the pulse parameters for any given type. Some manufacturers therefore pre-program their power sources so that the appropriate wire type can be selected from a menu.

In the past, wire feeders were often the Achilles' heel of equipment for welding with tubular wire. Feeders designed for solid wire have smooth drive rolls, one of each pair being grooved to provide the wedging action which generates the frictional force for feeding. High pressures are unlikely to damage the wire. The major cause of feeding problems with tubular wire was over-tightening of the drive rolls. This, especially with early wires which were often softer than today's, could squash the wire, increasing the risk of buckling and allowing powder to escape from the seam and clog the conduit liner. Modern feeders often have twin drive rolls to increase the tractive effort without increasing the pressure on the wire. For optimum efficiency, the pairs of rolls should either be independently driven or should be linked by differential gearing to prevent slip caused by production tolerances on roll geometries.

For wires of more than 1.2 mm diameter, knurled rolls are often used. Particular care is again needed not to tighten the rolls so much that they damage the wire surface, causing it in turn to wear the contact tip unnecessarily quickly.

2.3 Benefits

When tubular wires were first reintroduced in the 1950s, they were competing chiefly with manual metal-arc electrodes and before even considering deposition rates, the increased duty cycle of a continuous process was a major benefit. The first tubular wires thus replaced stick electrodes, although small diameter wires were not at first available and only downhand welding was possible. Later, the GMAW process with solid wires was widely promoted and the cost of the wires fell, especially in Europe. Users converted from MMA to GMAW and a more sophisticated case then had to be made for tubular wires.

The widespread acceptance, in Europe at least, of tubular wires was not helped by a misunderstanding by those promoting their use of where their real benefits lay. A fact easily grasped by marketing departments was that at a given current, a tubular wire will generally deposit metal at a faster rate than a solid wire of the same diameter. This is because almost no current is carried by the core of a tubular wire, even in the case of metal-cored wires. The smaller cross-section of the current-carrying sheath leads to more resistive heating in the electrode extension and a faster burn-off rate. Fabricators were often ahead of salesmen in realising that this is not unambiguously beneficial.

There is no reason why consumables should be compared at a fixed current, unless this is limited by the welding equipment. The current used in a particular application should be determined by the productivity required, the properties of the weld metal and heat-affected zone, and the geometry and appearance of the weld bead. Flux-cored wires have been used in semi-automatic applications at 600 A: in such cases, the use of a wire designed for CO_2 shielding reduced the radiation and ozone levels which would have made the use of a solid wire unpleasant, and a rather stiff slag maintained a good fillet weld profile. The same 2.4 mm wire has been used in a mechanised application at 900 A, depositing about 19 kg/h of weld metal.

In recent years, among the largest users of tubular wires have been shipyards. Here, metal-cored wires have been used in place of solid wires for the fillet welding of stiffeners and box sections. While similar deposition rates have been claimed for solid wires used at very high current densities, the driving force for the use of the more expensive tubular wires has been the good penetration, underbead shape and surface profile of the welds, resulting in very low defect levels. Paint manufacturers and welding consumable manufacturers working together have also been able to develop combinations of prefabrication primers and tubular wires that minimise problems when welding over primer.

In offshore construction, rutile tubular wires are used in positional welding at deposition rates much higher than those of either covered electrodes or GMAW with solid wire. Here, the slag is stiff enough to allow welding uphill in a free-flight transfer mode at currents up to 250 A, giving deposition rates above 4 kg/h. At the same time, metallurgical control which would not be possible with a solid wire ensures the required combination of strength and toughness.

In Europe, tubular wires still only constitute about 10% of total welding consumable consumption, while in Japan and North America the percentage is 35 or more.[4] In a Korean shipyard, the use of tubular wires is quoted as 94% of the total. It has at times seemed that the adoption of tubular wire in Europe was fuelled more by the knowledge of what international competitors were doing than by a thorough understanding of process costs and benefits. However, in all markets the proportion of welding using tubular wires is foreseen to rise in the immediate future,[4] and that consensus is a good indication that the potential of the process is no longer in doubt.

2.4   Materials used in tubular cored wire welding

Early flux-cored wires were developed by covered electrode developers who simply took formulations from MMA consumables and turned the product inside out. However, they failed to capitalise on the great potential advantage of the tubular form: while MMA electrodes need a binder to stick the powder to the core wire, in tubular wires the powder can usually be held within the sheath without the use of binders. Since the binders, alkali metal silicates, are the main source of moisture pickup in MMA electrodes, it might have been expected that their removal from flux-cored wire would be a first priority, but instead they lingered on for many years as arc stabilisers. Flux-cored wires gained a reputation for high hydrogen contents and porous welds which they did not live down until the 1980s.

2.4.1   Basic wires

Basic MMA electrodes were based on a simple lime–fluorspar–silica ($CaCO_3$–CaF_2–SiO_2) flux system, which translated easily to tubular wires. It was only necessary to lower the lime-to-fluorspar ratio to prevent excessive evolution of CO_2 as the wire was heated in the electrode extension, which would cause the wire seam to blow open towards the tip. Basic slag systems produce weld metals low in oxygen, which is good for ductility and toughness. Unfortunately, oxidation of the surface of the transferring droplet is the best way to lower its surface tension so that fine droplets become stable: since this does not happen with basic wires, the metal transfer tends to be quite globular. The use of electrode negative polarity is recommended for many basic wires as this can give a finer droplet transfer at low currents. As with basic MMA electrodes, the fluorspar inhibits hydrogen pickup by the droplets, so early basic flux-cored wires gave weld metals much lower in hydrogen than those from rutile wires. In addition, given that the metal transfer would never be very smooth, developers tended to forego the use of alkali metal-based arc stabilisers, so producing relatively non-hygroscopic wires.

The latest rutile wires come very close to matching basic wires in many aspects of their performance, including weld metal hydrogen levels, but in one aspect they are left behind. The low oxygen basic weld metals have the potential to give good toughness at much higher strength levels than rutile types do. As the use of steels with yield strengths exceeding 700 MPa increases, and especially as the use of strain-based design requires the weld metal strength to overmatch that of the parent material, basic wires may be the only way to achieve satisfactory mechanical properties. This will certainly be a challenge to consumable developers, since lime–fluorspar slags are inherently fluid and difficult to manage in positional welding, while the globular transfer leads to some spatter. Help may be at hand in the form of advanced power sources: it has even proved possible in the laboratory to weld X100 pipes in the fixed position with a basic wire giving a yield strength of more than 800 MPa, using pulsed arc welding.

Other areas where basic wires have traditionally been used rely on the slag fluidity to allow gases to escape and minimise porosity. Thus when welding over surfaces contaminated with oil or grease, or with thick primer coatings, basic wires may still be the best choice.

2.4.2 Rutile wires

Rutile flux-cored wires have suffered by association with rutile stick electrodes, which are perceived as easy to use but high in hydrogen and unable to deliver a high level of mechanical properties. However, designers of gas-shielded tubular wires have a freer hand – unlike rutile MMA electrode designers, for example, they do not have to rely on steam as a shielding medium. Over the years, a great variety of rutile flux-cored wire formulations has been tried, but the defining characteristic has been the ability of the wire to give extremely smooth, free flight metal transfer over a wide range of currents.

As discussed above, the factor which above all controls the size of the transferring droplets is their oxygen level. It is quite possible to formulate a rutile wire which will give a low oxygen weld metal, but it then reverts to the globular transfer characteristic of basic types. In moving from low strength rutile wires to wires giving up to 700 MPa yield strength, designers might lower the weld oxygen level from 650 to 550 ppm, but any further decrease would be likely to lead to unacceptable welding characteristics. At this strength level, the toughness obtained with rutile wires cannot therefore match that from basic wires, whose weld metals would have oxygen levels of 450 ppm or less. Nevertheless, a considerable metallurgical feat was achieved by a number of consumable manufacturers in the 1980s when they produced rutile wires that reached the levels of charpy and crack tip opening displacement

(CTOD) toughness demanded for offshore platforms for the North Sea. The use of microalloying with titanium and boron, although not always apparent in published product specifications, played a large part in this.

Another characteristic contributing to the versatility of rutile slag systems is their ability to develop a range of melting points and viscosities. Rutile melts between 1700 and 1800 °C, so with the addition of suitable fluxing agents it is easy to make slags with melting points around 1200 °C and to fine tune these according to the application. Thus, for uphill welding where the slag has to support the weld metal and mould it to a flat contour, rutile wires are pre-eminent. With some wires, currents up to 300 A can be used for vertical-up welding without losing control of the pool.

For high current downhand fillet welding, on the other hand, a slower freezing but more viscous slag may give the best results and this too can be readily formulated on a rutile base. Many such wires were in the past designed to run with CO_2 shielding, since this allows cooler running of the torch and is more comfortable for the operator, but where mechanisation or automation is possible, there is a tendency to replace them with metal-cored wires running on gas mixtures rich in argon.

A survey of rutile flux-cored wires on the British market in 1968[5] found weld metal hydrogen contents up to 31 ml/100 g. Those giving the highest values contained significant amounts of hygroscopic synthetic titanates and were produced by a drawing process using solid soap which was left on the wire surface. Already, other wires in the survey, using better formulations and a soap-free production route, were giving less then 10 ml/100 g of deposited metal hydrogen and pointing the way to today's figures of less than 5 ml/100 g for many wires. Rutile flux-cored wires are easy to use and are available for many types of steel, from mild steel to high strength and creep-resisting steels. It is therefore not surprising that they have overtaken basic types in popularity in the last 20 years, and growth in their use is expected to continue.

2.4.3 Metal-cored wires

Metal-cored wires were patented in 1957,[6] mainly as a means of overcoming a current shortage of solid welding wire. A later patent of 1974[3] described the addition of small amounts of non-metallic material to the core together with the metal powders, and mentions some of the features that have subsequently made metal-cored wires so useful in their own right.

While a high deposition rate per ampere may not in itself be a deciding factor in wire selection, metal-cored wires add to this a good underbead profile on argon-rich gases, which reduces defect incidence. This makes them particularly suited to fillet welding, as in shipyards and in the manufacture of earth-moving equipment. Slag levels are low, so it is possible to make three or more runs without deslagging. The lack of slag also makes welding over primers easier.

Like solid wires, metal-cored wires must be used in the dip transfer mode when welding uphill, which means that deposition rates are limited and they may not offer much advantage over solid wire. However, where vertical-down welding is permitted, very high deposition rates can be achieved: for example, a 1.2 mm wire running at 280 A can deposit 5.5 kg/h. Another approach to positional welding is the use of a pulsed arc. This allows a deposition rate of about 2.3 kg/h in the uphill direction and is very suitable for robotic applications where guidance is by through-arc sensing.

Mechanisation allows metal-cored wires to be used at high productivity and the recent commercial development of tandem pulsed arc welding takes this a step further. In this system, the wires are fed through contact tips which are insulated from each other but share a gas shroud. If the pulses on the wires are 180° out of phase, the arcs do not interfere with each other, although other methods are also possible. This system has proved very effective in shipbuilding and is now starting to be used for the circumferential welding of pipelines.

The introduction of metal-cored wires in Europe coincided with a period of active involvement of the gas companies in the welding field and was used by them to promote the sales of argon-rich gases. The 1974 patent,[3] which mentioned that metal-cored wires may be formulated to run well under CO_2 shielding, was forgotten for a number of years. However, in Japan, the welding consumables industry remained independent of gas suppliers and such wires became popular. Now that less of the European welding industry belongs to gas companies, metal-cored wires for use under CO_2 are available from European manufacturers as well.

2.4.4 Self-shielded wires

Although self-shielded tubular wires of large diameter were made in Austria before World War II,[2] the pedigree of products currently on the market is generally considered to date back to the late 1950s, when new wires were announced almost simultaneously in the USA[7] and the USSR.[8] The driving force was the need for a product which would be faster to use than stick electrodes, but which would be independent of a supply of shielding gas. The latter is important not only in areas where the infrastructure for supply does not exist, but also, for example, in welding on tall structures where heavy gas bottles could be hazardous.

Arc welding requires the pool and transferring metal to be protected from atmospheric contamination by oxygen and nitrogen, which cause porosity and degrade the mechanical properties of the weld metal. In the absence of an external shielding gas, it was found that the lime–fluorspar system borrowed from basic stick electrodes produced a gas shield if the wire stickout was long enough for the heat generated to break down the lime into CO_2 and CaO, while the fluorspar vaporised in the arc. The addition of metallic aluminium, as a deoxidant and nitride former, allowed more normal stickouts to be used and a range of wires was produced which could be used without external shielding gas.

Early self-shielded wires were not remarkable for their mechanical properties, mainly because the shielding did not altogether exclude nitrogen but trapped much of it as aluminium nitride in the weld metal, while the excess aluminium remained in solution in the weld metal. Aluminium is a strong ferrite former and if sufficient is present to reduce austenite formation, the beneficial austenite transformation products that give steel its combination of strength and toughness may not be formed. Much ingenuity went into developing other shielding mechanisms, involving for example lithium compounds which produce metallic lithium in the arc, so that aluminium levels could be reduced and the toughness improved. So successful were these efforts that by the mid-1970s, self-shielded wires were being used to weld thick section offshore platforms for the Forties Field in the North Sea, meeting stringent charpy and CTOD requirements.

A particular advantage of self-shielded wires in that application, as in the welding of high-rise buildings, was their relative immunity to winds and draughts. This is because the metal vapour shielding is not easily blown away and the weld is still rich in nitride formers and deoxidants.

Over the years, self-shielded wires were developed for many different applications, from high deposition rate welding of heavy plate to the welding of thin sheet at low currents and voltages. Special wires have been made for semi-automatic welding of pipeline girth welds, for welding galvanised steel and for 'gasless electrogas' welding. All these are excellent products and to try them, or to read the patents which describe them, is to be impressed by the way in which physical, chemical and metallurgical challenges have been overcome to produce them. Nevertheless, the materials which provide the shielding need significant amounts of energy to melt or vaporise them so that they can do their work, and this reduces deposition rates compared with gas-shielded wires. Moreover, the metal vapours eventually condense to form either particulate fume or condensates on any exposed surfaces around the weld. The high level of deoxidation increases the surface tension of the droplets, making it difficult to achieve fine droplet transfer. The use of rutile to control slag viscosity is ruled out in products that aim to have high toughness because in the presence of strong deoxidants, metallic titanium is reduced into the weld metal, where it leads to severe secondary hardening and embrittlement by forming titanium nitrides.

For these reasons, self-shielded wires tend not to be the consumables of choice where the option of a gas-shielded product is open. A further reason often cited is the use, in many self-shielded wires, of barium compounds. These have low work functions and reduce the voltage drop at the arc cathode, allowing barium-containing wires to operate at voltages up to 8 V less than their barium-free counterparts. This reduces nitrogen pickup and, together with the very low latent heat of fusion of barium compounds, increases the welder's feeling of control in positional welding. Unfortunately, some barium compounds are toxic, and although those used in self-shielded wires are insoluble and there is no epidemiological evidence of harm to welders' health, there are concerns about whether macroscopically insoluble compounds might be absorbed by the welder if present as sufficiently fine particulates in the fume.

There is certainly a place for self-shielded wires. They are convenient to use, particularly suitable for outdoor applications and torches are lightweight. Indeed new self-shielded wires are still actively being developed. However, users concerned with the highest productivity may conclude that gas-shielded consumables have more to offer.

2.4.5 Wires for stainless steels

Tubular wires for welding stainless steels may be of any of the types described above, but have some special characteristics of their own. Early wires were often made with basic slag systems, but since neither hydrogen cracking nor toughness is a problem with most stainless steels, attention soon moved to rutile types, which promise great ease of use and excellent weld appearance. The lower melting point of stainless weld metal compared with low alloy types means that for positional welding, more support from the slag is needed. This means that with stainless flux-cored wires, the difference between the all-positional types with fast freezing slags and the downhand types with more fluid ones is more marked than it is with mild steel wires and it is worthwhile to choose the right wire for the application.

Many stainless flux-cored wires are of Japanese origin and were designed for CO_2 shielding. Users should be aware that while these are capable, where required, of depositing weld metals with less than 0.04 % C, for lower levels an argon-rich shielding gas is likely to prove more reliable. Metal-cored stainless wires are available and are especially suited to use with pulsed power sources, provided these are suitably programmed – programs for mild steel wires will not be optimised for stainless ones. It then becomes possible to weld positionally at high speed and to make small fillet welds with a mitred profile that would be difficult to achieve by any other means. Despite this, efforts to sell such wires in Europe have generally been unsuccessful.

It is relatively easy to make self-shielded stainless flux-cored wires because their high chromium content increases the solubility of nitrogen in the weld metal and reduces the risk of nitrogen-induced porosity. However, the inferior handling characteristics of such wires compared to the gas-shielded types has led to a gradual decline in their popularity and they are now mainly used in the hardfacing, repair and maintenance sectors.

2.4.6 Wires for submerged arc welding

Although tubular wires for submerged arc welding have been on the market for more than 30 years, and the benefits claimed for them today were being discussed from the outset, they made very little impact until the 1990s. As with gas-shielded tubular wires, the deposition rate in submerged arc welding increases when a tubular wire is substituted for a solid wire at the same welding parameters, typically by 20–25%. In multipass welds, this benefit comes with no penalty other than the extra cost of the wire itself.

An interesting option, discussed for 30 years but only now achieving commercial success, is the possibility of using relatively acid submerged arc flux in combination with a basic tubular wire. Acid fluxes give a good bead appearance and slag detachability, and are less susceptible to moisture pickup than basic fluxes, but do not generally produce the tough welds needed, for example, in offshore applications. However, if used with a basic flux-cored wire, all the good attributes of the flux are retained, while the small amount of basic components, delivered directly to the arc cavity, lower the weld oxygen level and allow good toughness to be achieved. The process is shown in Fig. 2.2. The combination of acid and fused fluxes with basic tubular wires has now been taken up by the offshore industry and has many potential applications in other sectors.

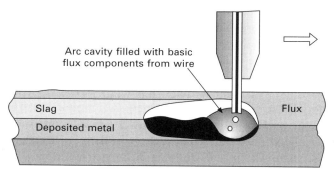

2.2 Using a basic tubular wire with an acid flux in submerged arc welding.

2.4.7 Manufacturing methods

Users of tubular wires will not always need to know the details of how they are manufactured, but a brief explanation may help to show why wires

32   New developments in advanced welding

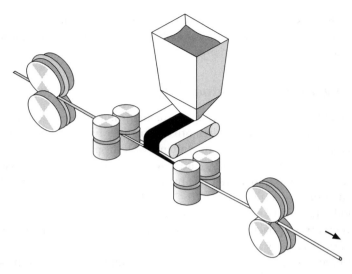

2.3 Line for forming and filling tubular wire.

sometimes look and behave differently. Most wires are made by starting with a flat steel strip and rolling it to form a tube. At the point where it is U-shaped, powder is poured into it (Fig. 2.3). After the tube has been closed, which is typically at a diameter of a few millimetres, there are different ways of reducing the wire to its final diameter. It may be drawn through dies, like solid wires. In that case, the dies must be lubricated and special drawing soaps are available for the purpose. However, these compounds contain hydrogen and to prevent this finding its way into the weld, the wires have to be baked to remove the organic components of the soap by oxidation, ideally leaving a non-hygroscopic residue. By careful selection of the soap, this residue can also be made to act as a cathode stabiliser, allowing the wire to operate with electrode negative when needed.

Alternatively, the wire diameter may be reduced by rolling. This requires much less lubrication, so no baking is needed after production. This results in a cleaner wire surface which gives a lower electrical resistance at the contact tip, so the voltage delivered to the wire is more constant and the amount of arc stabilisers in the wire can be reduced, further lowering its potential hydrogen content.

It is possible to make seamless tubular wires by starting with a tube of 12 mm or so in diameter and 15 m in length, and filling it from one end. A series of filled tubes are then joined end-to-end before being drawn down to the final diameter. Because this typically involves a 10:1 reduction of diameter, an intermediate heat treatment at a high temperature is needed and this restricts the use of reactive materials, for example some deoxidants, in the core, which the developer of seamed wires might use. However, seamless wires can be coppered for lower contact resistance and should not pick up moisture during storage.

2.5 Optimising productivity

Most of the guidelines that apply to increasing productivity for other welding processes apply equally when the consumable is a tubular wire – fillets are more effective than butt welds, downhand welding is faster than positional welding, roots made on a backing are faster than open roots and so on. The wide range of tubular wires available makes it easier to optimise productivity in any situation.

For downhand welding, large diameter rutile wires can be used to give deposition rates approaching 20 kg/h, or similar rates can be achieved using the twin arc process, where the wires are connected in parallel so only one power source is needed. As described above, tandem welding offers greater versatility at some increase in equipment cost. Where downhand welding is not possible, welding downhill with a metal-cored wire can deposit 5.5 kg/h or welding uphill with a rutile wire, 4 kg/h. The latter are figures that would be difficult to match with any other process. Because tubular wires, even of the flux-cored type, produce less slag than MMA electrodes, it is possible to use narrower joint preparations without running the risk of trapping slag: so where, for example, a joint with a 60° included angle might be used with MMA electrodes, 50° might be used with rutile flux-cored wire and 40° with a metal-cored wire. In mechanised pipe welding, metal-cored wire has been used downhill in preparations as narrow as 6° and was found to fill the joint with 20% fewer runs than a solid wire.

By choosing a suitable rutile wire, large standing fillet welds can be made in a single pass or, with metal-cored wires, a fillet of several runs can be made without deslagging. Basic and metal-cored wires can be used on an open weld root with no backing, but as is the case when unbacked roots are used with other processes, this must be done with a low current to avoid burning off the joint edges and the process is therefore slow. However, rutile wires perform very well with a ceramic backing, which allows relatively high currents to be used and is highly productive.

Tubular wires lend themselves particularly well to mechanisation and robotisation. In the first place, these move the operator out of the weld zone with its heat, radiation and fume, allowing welding conditions to be used routinely that would be prohibitively uncomfortable for a welder. Secondly, only mechanised processes can take advantage of the very high welding speeds of which tubular wires are capable. In addition, with wire regularly available in packs of 300 kg or more, the duty cycle can be increased to a high level.

2.6 Process control and quality

Tubular wires are capable of operation over a much wider range of conditions than many other processes. For example, a basic or metal-cored wire might be used at less than 100 A in positional welding with dip transfer, or at over 300 A in downhand welding. At these extremes, users might need to consider the implications of a low or high heat input, while at some intermediate currents and voltages, globular transfer might threaten poor bead appearance, lack of fusion and increased spatter. It is therefore important that despite the apparent ease with which untrained operators can pick up a torch and start welding with tubular wire, proper training is given in setting up the equipment and selecting appropriate parameters.

None of the problems that can occur with tubular wires are unique to the process and welding engineers will be familiar with their causes. However, because the parameters may change by a large factor with few visible signs that anything is different, it is more important than with other processes to be able to monitor parameters in high integrity joints. Especially when welding high strength steels, the properties of both the weld metal and the heat-affected zone are strongly affected by heat input. Equipment is now widely available to keep a continuous record of all welding operations and this is becoming standard in critical applications.

Gas metal-arc welding with solid wire has failed to gain acceptance in some areas because a perfect weld surface can conceal serious lack-of-fusion defects, which may be difficult to detect by radiography. Even with solid wire, the problem is being overcome with improved power sources and the increasing use of ultrasonic testing, but tubular wires offer a further reduction in susceptibility. Their better arc profile, especially when welding with argon-rich gases, and wettability have allowed them access to applications hitherto closed to gas-shielded welding. In vertical-up joints, the ability of rutile flux-cored wires to operate at twice the current of solid wires has led to their near monopoly of positional welding in offshore fabrication. However, in horizontal–vertical butt joints, where often only the upper plate is bevelled, positional rutile wires may not be best because their stiff slag is not needed to support the pool and may become trapped: in this case, a basic wire is usually preferred.

2.7 Applications

It has been claimed with some justification that there are now no 'no-go areas' for tubular wire. Even the power generation and pressure vessel industries, formerly seen as bastions of conservatism, have adopted them. The sector using the greatest quantity of tubular wires is shipbuilding, where the large amount of mechanised fillet welding to be done is ideal for the tubular wire process. The ability to weld over prefabrication primer with metal-cored or low-slag rutile wires is also important here.

Offshore construction is another high tonnage user of tubular wires, since many of the joints have to be welded in position and no other process can do this so productively. In covered yards, nickel-containing rutile wires are most productive, but for final assembly where conditions may be windy, self-shielded wires are often used.

Manufacturers of earth-moving equipment pioneered the use of flux-cored wires in the USA and introduced it into their European factories in the 1960s. This is a highly competitive and innovative sector where there has been heavy investment recently in laser welding, but tubular wires are still in a leading position. The use of metal-cored wires with robots using through-arc sensing for guidance will quite possibly be a productive technology in the future.

Flux-cored wires were used on German submarines during World War II[9] and later this was one of the first examples of their use on high strength steels. Crane jibs and offshore jack-up rigs were other examples where steel yield strengths up to 690 MPa were welded. The potential of the process for pipeline welding has yet to be fully exploited, but as X80 pipelines with 550 MPa yield strength have started to become more widespread, the use of tubular wires to weld them is increasing.

2.8 Troubleshooting

Improved quality control by all manufacturers over the last 20 years has meant that problems in using tubular wire have been much reduced.

2.8.1 Arc instability and feeding problems

Welders occasionally report what they describe as arc instability and snatching or sticking of the wire in the torch. The difficulty is to know whether electrical or mechanical problems came first: poor electrical contact can cause arc stability and an increase in feeding force as the wire momentarily welds itself to the tip, while feeding problems caused by a kink in the wire can result in similar voltage fluctuations and instability. Consumable manufacturers use high speed, multi-channel recorders to reveal the order of events, but the welder cannot always be expected to know which came first. The user is advised to check the equipment to see that the conduit and contact tip are of the right size and in good condition and that there is no build-up in them of debris from the wire surface. Feed rolls should not be over-tightened, as this can lead to powder escaping from the wire seam and causing clogging. The wire should be free from kinks and its curvature should not be excessive. There should be a small amount of lubricant on the wire surface. This may appear to vary from manufacturer to manufacturer and if there is any doubt about what is the correct amount for a given wire, the supplier should be consulted.

2.8.2 Porosity

Although porosity tended to be a fact of life in tubular wire welding in the 1960s, the lowering of hydrogen levels since then has made it normally easy to avoid. Hydrogen is the commonest cause of porosity in steel weld metals, arising in the past from excessive drawing lubricant on the wire surface or from hygroscopic materials in the core, but today more often from contamination of the plate surface by paint, grease or rust. The wires most susceptible to this are the all-positional rutile types, because of their stiff slags – basic wires, with their fluid slags, and metal-cored wires, with almost no slag, are much more resistant. More oxidising shielding gases, especially pure CO_2, help to prevent hydrogen absorption by the droplet and so reduce porosity.

In the past, tubular and solid wires with increased deoxidant levels were often used when welding on oxidised plates. However, they were not widely used in Europe, where the argument has been that if the oxide surface contains rust, the hydrogen in this is at least as likely to be a cause of porosity. Oxygen is more likely to cause porosity when it is entrained from the atmosphere together with nitrogen. Manufacturers recommend what gas flow rate to use for a given current and nozzle diameter. Problems can still occur either when welding in windy or draughty conditions, or if the gas shroud becomes clogged with spatter. Self-shielded wires rely on preheating of the core components as the wire moves from the contact tip to the arc to provide instant shielding and are prone to porosity if the stickout is too short.

Related to porosity is the appearance of gas trails on the surface of the weld. These happen when gas is trapped at the interface between the weld metal and the slag as they solidify. If the slag is relatively stiff, the gas cannot escape through it and moves horizontally to create a characteristic 'worm trail' on the weld surface. In a marginally more acceptable version, the gas does not move so far but leaves flat areas a few millimetres across on the weld surface. Both these types of defect are most likely to occur if a wire designed for all-positional welding is used at high current and with a low stickout in the downhand position.

2.9 Advantages and disadvantages

Welding processes are sometimes classified according to their power density. Those described in this book cover a wide range, with tubular wire welding somewhere in the middle. At the lower end of the scale, electroslag welding may run at less than 0.1 W/mm^3. Low power density processes such as this do not require accurate machining or alignment of the joint edges and do not subject the weld to rapid heating and cooling cycles that could cause problems with hardness. On the other hand, energy is being lost through conduction almost as fast as it is being applied to the weld, so the processes are thermally inefficient.

At the other end of the scale, lasers can concentrate power in a very small volume of material at 50 kW/mm^3 or more. Joint edges must be machined and the welding system must be capable of very accurate tracking so that it does not miss the joint. Rapid heating and cooling rates lead to the formation of hard martensite in the softest mild steel and the high welding speeds demand high purity base materials if hot cracking is to be avoided. The melting efficiency of power beams may be high, but until now most of this benefit has been lost because of the inefficiency with which the beams themselves were generated. One of the significant advantages of the high-energy density processes, however, is their ability to form a keyhole when welding onto a closed joint preparation, so allowing efficient welding from one side without backing.

Arc welding falls mid-way between these extremes, with power densities typically of the order of a few W/mm^3, and welding with tubular wires spans a useful range of the power density spectrum. This makes arc welding the most versatile of all the welding processes and tubular wires perhaps the most versatile of all consumables. They can be used semi-automatically with inexpensive equipment, as when small reels of self-shielded wire are used by hobbyists as an easy way of welding thin sheet. With more capital investment, shipyards can use them on tandem mechanised equipment at speeds which would need ultra-low sulphur and phosphorus levels for crack-free laser welding.

While most processes can be faster if joint preparations are accurate and consistent, this is not always possible, for example on the closing joints of large structures. In offshore fabrication, it is sometimes necessary to weld a brace to a chord with no internal access, a job which is not made easier by the acute angle between them. Here, self-shielded wires have been used to achieve excellent root profiles on the back of these difficult joints.

Another type of closing joint, however, shows the limitation of tubular wire welding in not being able to create a reliable keyhole. When the ends of two sections of pipeline meet and are to be joined, it is not possible to use an internal clamp and the fitup may be less than ideal. Cellulose electrodes can produce a root keyhole because hydrogen ions in their fierce arc give them good penetration and allow them to succeed in variable joint preparations. They are much faster than the self-shielded wires that would be a possible alternative. Cellulosic electrodes have therefore been widely used for the roots of pipeline tie-ins in steel grades up to X80 (550 MPa yield strength), although flux-cored wire is preferred for the filling and capping runs. But pipe welding procedures are well-proven to avoid hydrogen cracking: cellulose electrodes would not be an option for the offshore application described above where the section sizes are greater and the steels more hardenable.

Perhaps the greatest challenge for tubular wires remains solid wires, whose single but important advantage is consumable cost. This is not the same as the cost of the weld and there are many cases where a more expensive consumable reduces the cost of the job, but it is true that modern power sources can narrow the performance gap between solid and tubular wires. Nevertheless, the fact that the market share held by tubular wires is increasing throughout the world shows that the benefits of this versatile and productive process are far from exhausted.

2.10 Sources of further information and advice

Much of the information presented here is to be found in expanded form in the author's book *Tubular Wire Welding*,[10] while a companion book, *Self-shielded Arc Welding*[11] by T. Boniszewski, deals in more detail with that aspect of the process. More recently, the '*Flux Cored Arc Welding Handbook*[12] has been published. As with all welding processes, welding manufacturers are the primary source of information and several publish excellent handbooks and pamphlets describing tubular wire welding. They will also have technical specialists who can help with specific problems.

Research institutes such as TWI in the UK, the Edison Welding Institute in the USA and the Institut de Soudure in France are a key source of information for their members and are always working to solve the most intractable problems of the industry. Finally, all welders and welding engineers should be aware of the benefits of belonging to a professional body such as the Welding and Joining Society or the American Welding Society, whose members between them will have encountered and overcome just about every difficulty that an individual practitioner is likely to encounter.

2.11 References

1. British Patent Application 3762, Feb 14, 1911
2. Leitner F., 'Cored electrodes: their manufacture, properties and use', *Iron and Steel Institute Symposium on the Welding of Iron and Steel*, May 2nd and 3rd 1935, 115–30
3. British Patent 1510120, Nov 15, 1974
4. Tsutsumi S. and Ooyama S., 'Investigation on current usage and future trends of welding materials', IIW Doc. XII-1759-03, 2003
5. Salter G.R., 'CO_2 welding with flux-cored wires', *British Welding Journal*, 1968 **15**(5), 241–9
6. British Patent 858854, Mar 29, 1957

7. Wilson R.A., 'Vapor-shielded arc means faster welding', *Metal Progress*, October 1960
8. Pokhodnya I.K. and Suptel A.M., 'Mechanised open arc welding with cored electrodes', *Avtomatecheskaya Svarka*, 1959, **12**(11), 1–13
9. Rapatz F., 'Der heutige Stand des Seelendrahtes in der Schweißtechnik', *Electroschweißung*, 1943, **8**, 3–8
10. Widgery D.J., *Tubular Wire Welding*, Cambridge, UK, Abington Publishing, 1994
11. Boniszewski T., '*Self-shielded Arc Welding*', Cambridge, UK, Abington Publishing, 1992
12. Minnick W.H., '*Flux Cored Arc Welding Handbook*', Goodheart Wilcox Company, January 1999

3
Gas tungsten arc welding

B. L. JARVIS, CSIRO Manufacturing & Infrastructure Technology, Australia and M. TANAKA, Osaka University, Japan

3.1 Introduction

Gas tungsten arc welding (GTAW) first made its appearance in the USA in the late 1930s, where it was used for welding aluminium airframes. It was an extension of the carbon arc process, with tungsten replacing the carbon electrode. The new tungsten electrode, together with an inert helium shielding gas atmosphere, reduced weld metal contamination to the extent that highly reactive metals such as aluminium and magnesium could be welded successfully. For a time the process was known as 'heliarc' in the USA. Other countries substituted the less expensive argon for helium and referred to the process as 'argon-arc'. Later these distinctions were dropped and the process became known as tungsten inert gas (or TIG) welding. More recently the term gas tungsten arc (GTA) has been introduced to signify that the shielding gas may not necessarily be inert.

GTAW is known for its versatility and high joint quality. It can be used with a wide variety of materials, including highly reactive or refractory metals. It may be operated manually at lower currents (e.g. 50 to 200 A) for single pass joining of relatively thin sections, or multi-pass welding of thicker sections that have appropriate V- X- or similar type edge preparations.

During the 1960s the process was extended to much higher currents, allowing the arc forces to play a significant role in increasing weld penetration. At currents above about 250 A the arc tends to displace the weld pool, with the effect increasing as the current is increased further. This mode of operation is generally automated, and in its early manifestations gave rise to terms such as high current, buried arc, and sub-surface arc TIG (or GTAW). Plasma arc welding also has its origins in the GTAW process. More recent innovations have included the introduction of active fluxes (A-TIG), dual shield GTAW, guided GTAW, keyhole GTAW and laser-GTAW hybrid processes.

Understanding of the GTAW process involves input from many disciplines. Although appearing relatively simple, application of the process involves many choices including electrode size and composition, electrode tip geometry, power supply characteristics, electrode polarity, shielding gas, welding current, and voltage settings. Each of these will be related to the type of material and its joint geometry. The complexities and the importance of the GTAW process have stimulated research which is still very active more than 60 years after its introduction.

3.2 Principles

3.2.1 Energy transport

GTAW utilises an intense electric arc formed between a non-consumable tungsten electrode and the workpiece to generate controlled melting within the weld joint. Essentially the arc can be used as if it was an extraordinarily hot flame. The stability of the tungsten electrode and the option to use totally inert gas mixtures if desired means that the process can be very clean and easy to implement. It is also a process with the potential to deliver relatively high power densities to the workpiece, and so can be used on even the most refractory metals and alloys. It can be misleading to refer to arc temperatures as a measure of melting ability, but the intent can be captured in the measure of power density. Using this one finds that GTAW processes produce power densities at the weld pool of up to 100 W/mm^2. For comparison this is at least an order of magnitude greater than is available from an oxy-acetylene flame. The power density delivered to the workpiece is important in determining the process efficiency and can be a significant constraint when dealing with highly conductive metals such as copper.

Under standard conditions all shielding gases are extremely good electrical insulators. The current densities typical of welding arcs (of the order of tens of amps per square millimetre) can only be achieved if a high concentration of charged particles can be generated and maintained in the conducting channel. In arcs the necessary populations of electrons and ions are maintained by thermal ionisation and this requires temperatures of about 10 000K and above.

The degree of ionisation of a gas can be expressed as a function of temperature by the Saha equation (Lancaster, 1986). The resultant conductivity is then determined from consideration of the charge mobilities, as can be found in standard texts, e.g. Lorrain and Corson (1970) and Papoular (1965). An example of how the conductivity of argon varies with temperature is shown in Table 3.1 and presented graphically in Fig. 3.1. The data is taken from Lancaster (1986).

It is now known that the current density in an arc column has a limiting value under normal conditions. Once this limit is reached further increases in total current only distribute the current over larger areas of the anode, with

Table 3.1 Electrical conductivity data for argon between 3000 K and 30 000 K, at one atmosphere pressure (See Lancaster, 1986)

Temperature (degrees K)	Electrical conductivity (mho/m)
3000	0.00006
4000	0.127
5000	10.3
6000	101
7000	361
8000	923
9000	1770
10 000	2730
12 000	4740
14 000	6670
16 000	8200
18 000	9430
20 000	10 400
22 000	10 800
24 000	10 500
26 000	10 200
28 000	10 400
30 000	10 900

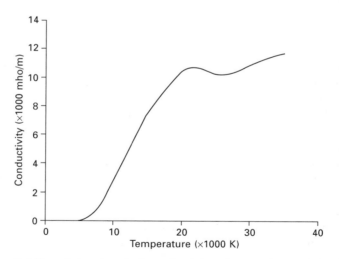

3.1 Plot of the electrical conductivity of argon from 3000K to 30 000K.

no appreciable change in peak current density on the arc axis (Jackson, 1960). In the case of argon the conductivity increases until doubly ionised argon appears at about 22 000 K. At this point the resistance provided by the doubly charged ions outweighs the benefit of the increased number of electrons

Gas tungsten arc welding

and so conductivity reaches a local maximum (see Fig. 3.1). Once this temperature has been reached in a particular region further increases in current will tend to generate a spreading of the current distribution into the adjacent, slightly cooler regions (Shaw, 1975).

For a very preliminary exploration of the welding arc, its main section can be treated as one-dimensional, i.e. as a function of radius, r, only. Such an approach, introduced by Glickstein in 1981, began with a simple model for the positive column in which ohmic heating was balanced against radial thermal conduction:

$$\sigma E^2 = \frac{-1}{r}\frac{\mathrm{d}}{\mathrm{d}r}\left(rk\frac{\mathrm{d}T}{\mathrm{d}r}\right)\mathrm{d}r \qquad [3.1]$$

In this equation σ is the electrical conductivity, E the electric field, r the radius from the arc axis, k the thermal conductivity and T the temperature. Equation [3.1] is known as the Elenbaas–Heller equation. This equation can be corrected for additional energy losses through radiation, $S(T)$, (Lancaster, 1986) and it is then known as the 'corrected Elenbaas–Heller', equation [3.2]:

$$\sigma E^2 = \frac{-1}{r}\frac{\mathrm{d}}{\mathrm{d}r}\left(rk\frac{\mathrm{d}T}{\mathrm{d}r}\right)\mathrm{d}r + S(T) \qquad [3.2]$$

Since the electrical and thermal conductivities of shielding gases have complicated temperature dependencies, as shown in Fig. 3.1, these equations can only be solved numerically. Nevertheless, the view of an arc in which radial conduction and radiation balance ohmic heating is easily visualised and so is useful in developing a qualitative understanding of arc behaviour. For example, Glickstein's solutions predicted that helium arcs should be much broader than those of argon despite peak temperatures and current density distributions being similar. Consequently, helium arcs should require higher voltages than argon arcs do – as is observed – since the energy is derived from the electric field. Similarly, it can be appreciated that vapour contamination or minor additions of a gas of lower ionisation potential should significantly alter the arc configuration.

An appreciation of the welding arc via the Elenbaas–Heller equation has two fundamental limitations: there is no consideration of the regions connecting the plasma to the electrodes and the omission of convection within the arc.

The very narrow regions between the electrode surfaces and the arc proper are known as sheath regions. In these regions the high temperatures (~10 000 K) needed for good electrical conductivity in the gas cannot be sustained due to the cooling provided by the cold electrodes (even the boiling temperature of iron is thousands of degrees below that required for argon to conduct well). Consequently, the electrical conductivity of the gas will be extremely low (see Table 3.1). Because of the high resistivity close to the electrodes the electric field of the arc will be very much stronger in these regions than elsewhere. This is equivalent to saying that the field has a non-zero divergence and according to Maxwell's equations must be associated with the presence of net electric charge:

$$\nabla \cdot E = \frac{\rho}{\varepsilon_0} \quad [3.3]$$

Or, in one dimension:

$$\rho = \varepsilon_0 \frac{dE}{dx} \quad [3.4]$$

Consequently, sheath regions will be bounded by regions of charge, one on the electrode surface and the other at the interface with the plasma. This latter constitutes a region of space charge. The corresponding voltage drops are sometimes known as 'fall' voltages (see Fig. 3.2).

The sheath regions are extremely important in determining the particular characteristics of an arc and in establishing the overall energy balance. In welding arcs the predominant charge carriers are electrons and these must be continually replenished by being drawn out of the cathode and across the cathode sheath. Liberating electrons from a metal surface requires a considerable amount of energy – each electron absorbing at least an amount $e\phi$ where ϕ is the work-function of the surface (typically 2–4 V). If the metal is suitably refractory (such as tungsten or hafnium) this can be provided by the high temperature of the electrode and is then known as thermionic emission. In this case the electrons effectively evaporate from the surface. If the temperature of the electrode is not high enough the electrons must gain their energy from the very high strength field between the surface and the surrounding space charge. This is termed field emission. GTAW is generally operated in the thermionic emission mode. Electron emission is aided by the presence of oxides and other surface impurities.

The electrons leave the arc by crossing the anode sheath and entering the anode, which is usually the workpiece. The anode sheath is believed to be of the order of one electron mean free path in width, to be consistent with its relatively low temperature. In crossing this and entering the anode the electron transports a considerable portion of the total energy flux. The energy contribution of each electron to the anode includes its thermal energy, the energy it absorbs from the anode fall, and its energy of condensation, $e\phi$. In some cases this can amount to as much as 80% of the total energy flow into the anode. The other major source of energy transport to the anode is conduction and here the characteristics of the shielding gas become important. For example, helium is far more conductive than is argon and consequently delivers more heat – hence the perception that it makes an arc 'hotter'. Gases such as

Gas tungsten arc welding 45

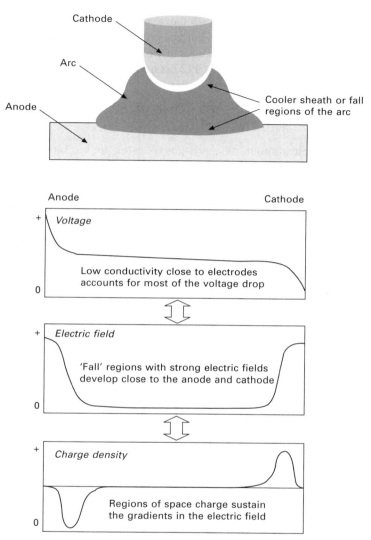

3.2 Schematic illustrations of the variation in voltage, electric field and charge densities with position along an arc discharge.

hydrogen and nitrogen exhibit what is known as 'reactive thermal conductivity'. They dissociate at high temperatures with the absorption of significant amounts of energy, only to recombine and release this energy in the cooler regions such as the anode sheath. So these gases are also associated with 'hot' arcs. In addition to electron absorption and conduction, convection and radiation also transport energy. Convection in particular becomes very important as currents rise above 40 A or so (Zhu *et al.*, 1992) and may be the dominating transport mechanism outside the sheath regions. Convective flow is powered by Lorentz forces associated with the passage of the high welding currents, and has an impact on the momentum as well as on energy transferred to the weld pool. Present numerical models of welding arcs endeavour to incorporate all these effects (Lowke et al., 1992, Zhu et al., 1992) but there is still much development to be done.

3.2.2 Momentum transport

At currents below about 200 A the gas tungsten arc has many characteristics of an ideal flame. It can be chemically inert, it produces very high heat fluxes to the workpiece, and it appears to produce almost no disturbance to the molten metal it produces. But despite the absence of metal transfer the arc does transport momentum and this becomes important at higher currents. The momentum transfer and several of the resultant forces on the weld pool are due to Lorentz forces generated within the arc. These forces can give rise to high velocity plasma jets. Similar forces also occur within the pool and are one of the drivers for circulation within it. The strength of these forces is dependent on the magnitude of the welding current ($F \propto I^2$) and its geometric distribution. The latter dependency is in turn related to variables such as electrode composition and geometry, and choice of gas shield composition.

In order to model a welding arc one might begin by considering the motion of an individual element of the plasma. Thus each element of the arc fluid is accelerated in proportion to the net force acting on it:

$$\rho \frac{dv}{dt} = \rho \left(\frac{\partial v}{\partial t} + v \cdot \nabla v \right) = \text{(net force per unit volume)} \qquad [3.5]$$

where ρ is the (incompressible) fluid density, v its velocity and t is time. The net force per unit volume in an arc will include the Lorentz term $J \times B$, the pressure gradient $-\nabla P$, and a 'diffusion' term that accounts for viscous damping, $\eta \nabla^2 v$. The resultant equation is a modified Navier–Stokes equation for an incompressible fluid, and reads:

$$\rho \frac{dv}{dt} = -\rho v \cdot \nabla v - \nabla P + J \times B + \eta \nabla^2 v \qquad [3.6]$$

Solving this equation for an arc is challenging since the parameters are strongly coupled, rendering the system non-linear. In general, several different equations must be satisfied simultaneously (e.g. conservation of mass, energy, charge and momentum) and numerical methods must be used for their solution. The work of Zhu, Lowke and Morrow (1992) provides a comprehensive treatment of this problem.

However, as is often the case, much can be learned by considering simplified approximations. One such approximation is to ignore viscosity, as did Converti

(1981). He treated the arc as a truncated cone with the welding current, I, flowing between the two electrodes, a tungsten tip with an emission area of cross-sectional radius R_e and the weld pool surface of larger radius R_a. With the assumption that the current density is constant over any chosen radial cross-section, the net force normal to the pool was found to be:

$$F = \frac{\mu I^2}{8\pi}\left(1 + 2\ln\frac{R_a}{R_e}\right) \quad [3.7]$$

The ratio R_a/R_e is known as the arc expansion ratio.

Converti identified the two $J \times B$ components contributing to the net Lorentz force acting on the arc. Current flowing through an arc generates a circumferential magnetic field, $B_\theta(r)$, perpendicular to both the axial and radial vectors. Consequently both axial and radial components of the arc current will interact with this field to give rise to forces. The axial component ($J_z \times B_\theta$) generates a compressive, or pinch force while any radial component ($J_r \times B_\theta$, due to arc expansion) results in an axially directed force. These two forces give rise to a radial pressure gradient and a fluid flow (the plasma jet), respectively. The radial pressure gradient produces a static pressure that squeezes the plasma against the terminating electrodes. On the other hand, the fluid flow contributes a dynamic pressure that acts only on surfaces that change the velocity of the fluid stream.

Evaluation of equation [3.7] indicates that the arc force increases with the square of the welding current. Furthermore, experimental observation (Erokhin, 1979) and calculations based on reasonable estimates of the arc expansion ratio (Jarvis, 2001) show that the magnitude is of the order of $3 \times 10^{-5} I^2$ grams weight. So for example, an arc carrying 100 A would exert a relatively insignificant force of about 300 mg weight, whereas at 500 A the force would be nearer 7.5 g weight. The latter is sufficient to displace a significant volume of weld metal, molten stainless steel having a density of about 7 g/cm^3.

Evidently the larger portion of the arc force derives from the dynamic pressure term $(\mu/4\pi) \ln (R_a/R_e)$. Consequently changing the arc expansion ratio will alter the arc force generated at a given current. Now, in the case where the tungsten electrode is the cathode there is good evidence that the emission current is approximately proportional to emission area. In fact, measured values for emission current densities vary slightly around about 150 A/mm^2, depending on electrode composition (Matsuda *et al.*, 1990) and welding current (Adonyi-Bucurdiu, 1989). Consequently the arc expansion ratio can be increased by measures such as reducing the angle of the electrode taper or changing the electrode composition. Other factors, such as choice of shielding gas and electrode diameter can also alter the expansion ratio by changing the thermal balance at either electrode (see also Section 3.4.6).

The arc pressure is a measure of the arc force per unit area at any given point over the weld pool. Generally arc pressure is a maximum on or close to the arc axis and is often modelled as having a Gaussian distribution. Arc pressure is sensitive to changes in the distribution of the arc force and so is significantly altered by factors such as redistribution of the current and changes in gas viscosity. For example, the arc pressures in a helium arc are significantly lower that those in an argon arc at the same current because high-temperature helium is more viscous than argon and therefore distributes the arc force over a wider area.

3.2.3 Weld pool behaviour

To complete a model of the GTAW process it is necessary to consider the behaviour of the liquid weld metal. The weld pool can be a very active part of the welding process, with significant energy and momentum transport taking place within it. In addition to Lorentz forces, the weld pool is subjected to variations in surface tension, buoyancy, marangoni and 'aerodynamic' plasma drag forces. Finally, at higher currents the pool surface can be highly distorted and this can modify current and gas flow within the arc, as well as produce another surface tension-based driver for the flow of the liquid metal (see below). In general, however, forces associated with gradients in surface tension are believed to dominate flow within the pool.

The flow resulting from gradients in surface tension is often referred to as marangoni flow (Lancaster, 1986). Normally surface tension decreases with increasing temperature, so that the weld pool surface will have a higher surface tension at the edges than at the centre. As a result the hotter weld metal at the centre is drawn across the surface to the edges, thereby establishing a circulation that transports heat directly to the edges of the pool, favouring the formation of a wide, shallow weld puddle. Under appropriate conditions this effect can be reversed by surface-active elements such as sulphur, phosphorus and selenium. These elements lower the surface tension in the cooler regions of molten metal, but are dissipated at higher temperatures. In such circumstances the temperature coefficient of surface tension can become positive (that is, surface tension could increase with temperature) and reverse the expected direction of flow. This circulation transports heat to the bottom of the pool rather than to the edges, to produce deep, narrow weld pools. In this way the performance of specific welding procedures can be compromised by heat-to-heat variations in sulphur content within a given type of stainless steel, for example. Lorentz forces also promote 'centre-down' circulation within the pool (see Fig. 3.3 and the discussion in Section 3.3).

When the arc current exceeds about 150 A the weld pool surface becomes noticeably concave in response to the arc forces. The degree of metal displacement increases with increasing current and becomes an important

Gas tungsten arc welding 49

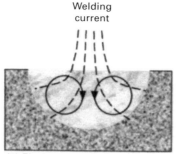

(1) Electromagnetic force: fluid driven by J X B forces.

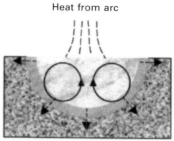

(2) Natural convection (buoyancy): hot fluid under the arc spreads, cooler fluid at the edges sinks to the bottom.

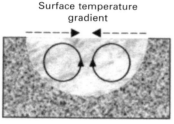

(3) Marangoni: surface is drawn by regions of highest surface tension (normally the coolest regions).

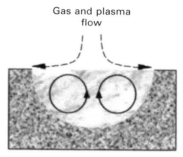

(4) Aerodynamic drag force: surface stress arises due to friction between surface and gas stream.

3.3 Flow directions induced by four possible motive forces in arc welding (Matsunawa, 1992).

influence on process performance above about 250 A. The displacement of the weld pool is visible as a terminating crater if the weld is abruptly terminated. Such craters are interesting for several reasons. For example, their presence indicates that the liquid displaced by the arc forces does not simply accumulate around the edges of the pool but actually gets frozen into the weld bead. The amount of material required to fill the crater has been referred to as the 'deficit' (Jarvis, 2001). Although the shape of the crater may differ from the depression of the pool during welding, it is evident that the deficit is conserved. Hence measurement of the deficit, via the terminating crater, can be used to provide useful insights into the weld pool dynamics.

The use of measurements of deficit is illustrated by the data presented in Table 3.2 and plotted in Fig. 3.4. The data is from experiments involving GTA bead-on-plate welds on stainless steel using alternately argon and helium shielding gas. What is evident in each case is an abrupt and large increase in

Table 3.2 Pool displacement produced under various welding conditions on 10 mm AISI 304 stainless steel plate

Weld speed (m/min)	Shielding gas	Plate thickness (mm)	Welding current (A)	Bead width (mm)	Pool displacement (g)
Melt-in	He	10	200	8.0	0.053
Melt-in	He	10	230	7.8	0.035
Melt-in	He	10	260	8.6	0.079
Melt-in	He	10	290	9.1	0.097
Melt-in	He	10	320	9.4	0.070
Melt-in	He	10	390	11.3	0.220
Melt-in	He	10	425	11.9	0.351
Melt-in	He	10	470	12.4	3.822
Melt-in	Ar	10	120	5.3	0.018
Melt-in	Ar	10	170	8.0	0.097
Melt-in	Ar	10	240	9.3	0.228
Melt-in	Ar	10	255	7.3	0.457
Melt-in	Ar	10	320	7.9	0.598

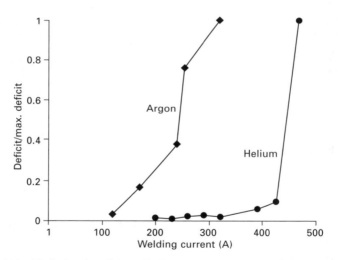

3.4 Variation in (dimensionless) deficit with current for melt-in mode GTAW.

deficit over small changes in current. These changes correspond to similarly large changes in penetration (Fig. 3.5). The implication is that an inadvertent choice of welding parameters near such transition regions could result in serious weld inconsistencies.

Models that balance arc forces against the combined effects of buoyancy and surface tension (Jarvis, 2001) could explain sudden changes in deficit.

Gas tungsten arc welding 51

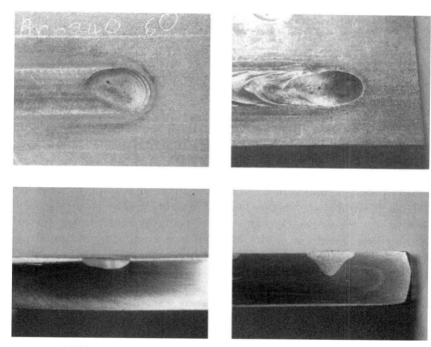

3.5 Visual evidence of abrupt changes in deficit for bead-on-plate welds on stainless steel. Both welds were made using argon shielding and at the same welding speed and voltage. Left 240 A, right 255 A.

Essentially the argument is as follows. If the width of a weld pool is fixed and the arc force is gradually increased from zero, surface distortion will be resisted by buoyancy and by surface tension. These forces increase as the curvature increases, hence deficit rises relatively slowly. However, the resistance provided by surface tension has a maximum value ($2\pi r \gamma$) that corresponds to the surface becoming vertical at some radius r. Further increase in arc force beyond this value causes proportionately much greater displacement as it is now only limited by the weaker buoyant forces.

Models that can describe weld pool surface geometry begin with the assumption of a 'free surface'. This means that the pool surface moves until the net pressure change across it is zero. Pressures arise from surface tension, buoyancy and arc pressure. Because the net pressure is zero everywhere the surface is at a local minimum in energy. Of course the surface is attached to the parent material at the boundary of the pool. It follows that when the boundary moves as the heat source moves along the joint, the distorted surface moves with it. (If it did not then its shape would change, its surface energy increase and it would experience a restoring force acting to realign it with the moved boundary). This automatically drives liquid metal from the leading to trailing edge of the pool and so is another potential driver for fluid flow within the pool.

3.3 The A-TIG process

3.3.1 Introduction

The tungsten inert gas (TIG or GTAW) welding process is suited to welding operations requiring considerable precision and high joint quality. However, these advantages are offset by the limited thickness of material that can be welded in a single pass and by the poor productivity of the process. The poor productivity results from a combination of relatively low welding speeds and the high number of passes required to fill the weld joints in thicker material.

A new process variant, known as 'A-TIG', uses an activating flux to overcome these limitations by increasing the penetration significantly that can be achieved at a given current (Lucas and Howse, 1996). The concept of using such a flux was first proposed by the E. O. Paton Institute of Electric Welding in the former Soviet Union (Lucas and Howse, 1996; Lucas, 2000; Howse and Lucas, 2000). The first published papers that described their use for welding titanium alloys appeared in the 1960s (Gurevich et al., 1965, Gurevich and Zamkov, 1966). The result is that the penetration depth can be dramatically increased, by between 1.5 to 2.5 times relative to the conventional TIG process, by the simple application of a thin coating of the flux to the surface of the base material before arcing (Lucas et al., 1996). Consequently, A-TIG is expected to bring about large productivity benefits and accordingly intense interest in this process has been shown recently.

3.3.2 Flux and equipment

A-TIG is a simple process variant that does not require any special equipment (Lucas and Howse, 1996). In contrast, attempting to increase the depth of weld penetration by using other processes such as plasma or laser welding would require a substantial investment in new equipment. Plasma welding requires specialised torches and power supplies, while laser welding is dependent on expensive, high power lasers and precision beam and component manipulation (Okazaki and Okaniwa, 2002). In addition, these processes can require a significant commitment to appropriate procedure development for specific applications. In contrast, the A-TIG process just needs conventional TIG equipment – a standard power source and TIG torch with the normal size and type of tungsten electrode. These items would be existing equipment in most workshops, laboratories, factories and plants (Okazaki and Okaniwa, 2002).

The activating flux is provided in the form of a fine powder which is

 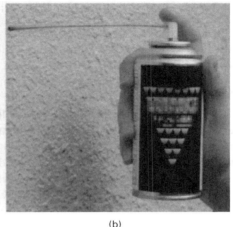

(a)   (b)

3.6 Simple techniques for applying the activating flux in the A-TIG process include (a) a brush and (b) a spray.

mixed with acetone or MEK (methyl ethyl ketone) into a paste and painted on the surface of the material to be welded (Lucas and Howse, 1996; Howse and Lucas, 2000; Tanaka 2002). The paste can be applied by a brush or with an aerosol applicator such as a spray (see Fig. 3.6). The A-TIG process can be used in both manual and mechanised welding operations (Lucas, 2000). The flux appears to be equally suitable for increasing the depth of penetration for welds produced with either argon or argon–helium shielding gases (Lucas and Howse, 1996; Anderson and Wiktorowicz, 1996).

Activating fluxes are available commercially from a number of companies in the UK, USA, Japan and so on (Lucas 2000; Tanaka 2002). There are many formulations which have been designed for welding materials such as carbon–manganese steel, low alloy steel, stainless steel, nickel-based alloy and titanium alloy. Although there are no published data formally on chemical compositions of commercial brands, there appears to be range of flux compositions in some literature (Ostrovskii *et al.*, 1977, Lucas and Howse, 1996; Lucas *et al.*, 1996; Ootsuki *et al.*, 2000; Tanaka 2002). The activating fluxes are predominantly composed of the oxides of titanium (TiO_2), silicon (SiO_2) and chromium (Cr_2O_3) with the addition of small quantities of halides as minor elements. Examples of included halides are sodium fluoride (NaF), calcium fluoride (CaF_2) and aluminium fluoride (AlF_3) (Lucas and Howse, 1996). As an example, the following flux composition has been reported for welding carbon–manganese steel and was produced in the former Soviet Union (Ostrovskii *et al.*, 1977): SiO_2 57.3%, NaF 6.4%, TiO_2 13.6%, Ti 13.6% and Cr_2O_3 9.1% (permissible deviation +/–2%).

3.3.3 Arc phenomena in the A-TIG process

It is well known that A-TIG shows a visible constriction of the arc compared with the more diffuse conventional TIG at the same current level (Lucas and Howse 1996; Lucas 2000; Howse and Lucas, 2000; Lucas et al., 1996).

Tanaka et al. (2000) have made and compared experimental observations of interactive phenomena between the arc and the weld pool in the A-TIG and conventional TIG processes. They employed pure TiO_2 as the flux, since a simple composition aided in the understanding of the phenomenon and this compound was one of the main elements of several fluxes on the market (Ostrovskii et al., 1977; Lucas and Howse, 1996; Lucas et al., 1996; Ootsuki et al., 2000; Tanaka, 2002). Figure 3.7 shows cross-sections of welds made with and without flux at the three different welding currents. The material was an austenitic stainless steel (AISI 304) of 10 mm thickness. The shielding gas was helium, the welding speed was 200 mm/min, and the arc gap was 5 mm. It can be seen from the figure that the depth/width ratio of welds with flux was higher than that of welds without flux, independent of welding current. This figure suggests that a satisfactory increase in penetration depth can be expected even with a flux consisting of only TiO_2.

The arc in the helium shielded TIG process has a characteristic appearance, both with and without flux. In the case without flux, there is a large, wide region of blue luminous plasma in the lower part of the arc. The blue luminous plasma appears to be mainly composed of metal vapour from the weld pool. In the case of A-TIG, the region of the blue luminous plasma is constricted at the centre in the lower part of the arc and the anode spot can be observed at the centre of the weld pool surface.

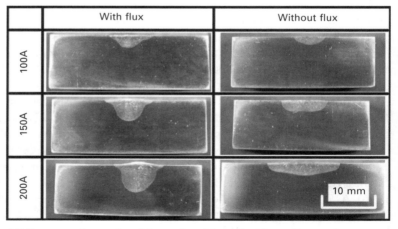

3.7 Cross-sections of welds made with and without flux at three different welding currents (Tanaka et al., 2000).

Figure 3.8 shows results of spectroscopic measurements of TIG welding arc plasmas with and without flux. Three line intensities, He I (438.793 nm), Cr I (425.435 nm) and Fe I (430.79 nm), were measured. Typical line spectra are shown in Fig. 3.9. Further line intensities of neutral metal atoms and metal ions could be detected, but line intensities of titanium (Ti I and Ti II) and oxygen (O I) could not be detected within the visible range. The intensity of each measured line is indicated by the grey scale in Fig. 3.8. In the case

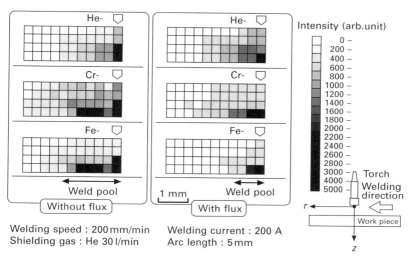

3.8 Spectroscopic measurements of arc plasmas in TIG welding with and without flux (Tanaka *et al.*, 2000).

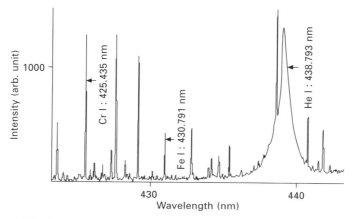

3.9 Typical line spectra for arc plasma in TIG welding process of type 304 stainless steel (Tanaka *et al.*, 2000).

without flux, the intense regions of Cr I and Fe I were considerably expanded in the lower part of the arc. However, in the case of A-TIG, both regions were only observed at the centre in the lower part of the arc. Therefore, it can be supposed that the blue luminous region is plasma mainly composed of metal vapour from the weld pool, namely the metal plasma. However, the He I region remained unchanged, i.e. it was independent of flux. This means that the arc constriction in the A-TIG process was associated with a change of metal vapour from the weld pool. It also appears that vapours from flux only weakly affect the arc constriction, as the line intensities of titanium and oxygen were very difficult to detect in the arc plasma.

It is well known that the metal vapour concentration depends on the surface temperature of the weld pool (Block-bolten and Eagar, 1984). Accordingly, pyrometric measurements of surface temperatures were included in the study by Tanaka *et al.* (2000). Figure 3.10 shows the radial temperature distributions on the weld pool surface with and without flux. Without flux, the surface temperature decreased gradually from the centre to the edge of the weld pool. With flux, however, the surface temperature at the centre of the weld pool was higher, at approximately 2350 K, while it became lower than the temperature without flux at an outer radius of about 1.5 mm. This means that the surface temperature gradient in A-TIG is much higher than in the conventional TIG process. It may be noted that a second peak in surface temperature appeared at about 2.5 mm radius for the A-TIG process (Fig. 3.10). However, this peak temperature was the surface temperature of flux heated directly by the arc on the unmelted base metal and so not related to the weld pool.

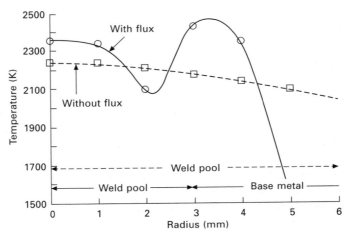

3.10 Radius temperature distributions on the weld pool surface with and without flux (Tanaka *et al.*, 2000).

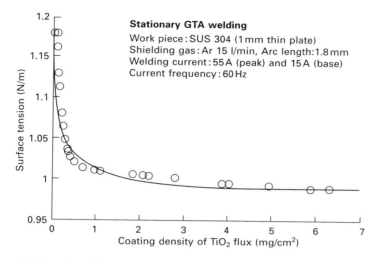

3.11 Relationship between surface tension and coating density of TiO$_2$ flux as determined by using weld pool oscillation (Tanaka *et al.*, 2000).

Tanaka *et al.* (2000) also measured the relationship between the surface tension and coating density of the TiO$_2$ flux by using the technique of weld pool oscillation (Xiao and den Ouden, 1990). It was found that at first surface tension decreased sharply with coating density but became approximately constant at densities greater than about 1 mg/cm^2 (see Fig. 3.11). This constant value of surface tension of about 1 N/m is quite similar to a value Ogino *et al.* (1983) measured by the sessile drop method at 1873 K. They studied the effects of oxygen and sulphur on the surface tension of molten iron and found that surface tension decreased sharply with oxygen content to a value of about 1 N/m at 300 ppm. Furthermore, they also showed that surface tension decreased equally sharply with sulphur content.

Tanaka *et al.* (2000) also investigated the relationship between the penetration depth and coating density of the TiO$_2$ flux, as shown in Fig. 3.12. They found that penetration depth increased sharply with the coating density before becoming approximately constant at densities greater than about 1 mg/cm^2. Using a standard technique, such as manual application by brush, the coating density of flux is approximately 15 mg/cm^2. The change in penetration depth in Fig. 3.12 correlates with the change in surface tension in Fig. 3.11. From these results, it may be concluded that surface tension is an important element in the mechanism of the A-TIG process.

Recently, Lu *et al.* (2002) have studied the effects of various oxide-based fluxes on the penetration depth in the A-TIG process. They selected five single component activating fluxes: pure Cu$_2$O, NiO, Cr$_2$O$_3$, SiO$_2$ and TiO$_2$,

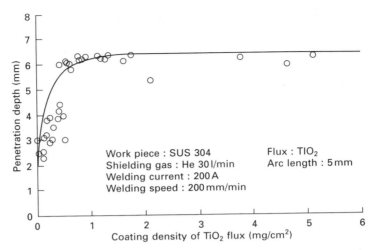

3.12 Relationship between penetration depth and coating density of TiO$_2$ flux (Tanaka *et al.*, 2000).

and investigated the relationships between the flux type, the depth/width ratio of the weld penetration and the oxygen content in the weld metal. Each flux gave a different depth/width ratio for each coating density, this being principally dependent on their relative chemical stability. However, Lu *et al.* (2002) found that the depth/width ratio increased by 1.5 to 2.0 times as the oxygen content in the weld metal passed through the range of 70–300 ppm, independent of the flux composition. Too low or too high oxygen content in the weld did not increase the depth/width ratio. They concluded that the oxygen from the decomposition of the flux in the weld pool altered the temperature coefficient of surface tension of the weld pool, which in turn changed the depth/width ratio of the weld penetration by inverting the direction of marangoni convective fluid flow. This is consistent with the results of Taimatsu *et al.* (1992) who showed that oxygen was an active element in pure liquid iron in the range of 150–350 ppm. They found that in this range the temperature coefficient of the surface tension of the Fe–O alloy was positive, while out of the range, the temperature coefficient was negative or nearly zero. Therefore Lu *et al.* (2002) clearly demonstrated that the oxygen from the decomposition of the flux in the weld pool was a key to the understanding of the A-TIG phenomena.

The presence of surface active impurities, such as oxygen, sulphur, etc., in the base material is known to affect the geometry of the weld bead (Makara *et al.*, 1977). These surface active elements also increase the penetration depth (Makara *et al.*, 1977). Katayama *et al.* (2001) directly observed the phenomenon of convective fluid flow in the weld pool by using a micro-focused X-ray transmission method during a TIG welding process. They

Gas tungsten arc welding

confirmed that different sulphur contents inverted the direction of convective flow in the weld pool of the same type 304 stainless steel. An outward fluid flow was observed for low sulphur content (40 ppm), whereas an inward fluid flow was observed when the sulphur content was high (110 ppm). It was confirmed that the difference in flow direction in the weld pool dramatically changed the geometry or depth/width ratio of weld penetration. In view of the above, it is concluded that the mechanism for the effect of the flux and that of surface active impurities in the base material are the same.

3.3.4 Review of process mechanism

Most researchers believe that arc constriction increases the current density and heat intensity at the anode root, enabling a substantial increase in penetration depth to be achieved (Howse and Lucas, 2000). A-TIG shows a visible constriction of the arc when compared with the more diffuse conventional TIG at the same current level (Simonik *et al.*, 1976; Ostrovskii *et al.*, 1977; Savitskii, 1979; Savitskii and Leskov, 1980; Howse and Lucas, 2000). It has been suggested that the arc constriction is produced by the effect of vaporised flux elements, namely oxygen or halogens (such as fluorine), capturing electrons in the outer (cooler) regions of the arc owing to their higher electron affinity as shown in Fig. 3.13 (Simonik *et al.*, 1976; Savitskii 1979; Howse and Lucas, 2000).

Savitskii and Leskov (1980) have proposed a further mechanism involving the interaction between the arc and the surface of the weld pool, because they could not observe arc constriction on a water cooled copper anode with

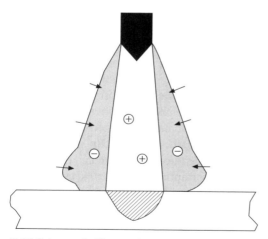

3.13 Schematic illustration of model of arc constriction by the activating flux (Howse and Lucas, 2000).

the flux. Since the surface curvature of the weld pool depends on a balance of arc pressure (cathode jet) and surface tension, the lower surface tension caused by the oxygen in the A-TIG flux should lead to greater surface curvature. This being the case, the cathode jet will be less able to dissipate metal vapour from the pool surface to the outer regions of the arc because the pool surface itself becomes an obstructive wall for the metal vapour, similar to a keyhole. As a result, the arc should be constricted owing to the concentration of metal vapour with low ionisation potential in the centre region of the arc. However, Savitskii and Leskov (1980) did not take account of the convective flow in the weld pool, which is described below.

Ostrovskii *et al.* (1977) argued that the main mechanism was a recirculatory flow driven by the electromagnetic force (the Lorentz force) resulting from the increase in current density at the anode root. He believed that the strength of the cathode jet should decrease and so not cause greater surface curvature of the weld pool, as stated in Savitskii and Leskov (1980), because the pressure difference between the cathode and anode would become very small as a result of the arc constriction. In fact, the keyhole phenomenon is not observed in the A-TIG process.

Heiple and Roper (1981, 1982) proposed that the change in the magnitude and direction of surface tension gradients at the weld pool surface caused by surface active elements such as oxygen, sulphur, selenium, etc. should change the direction of a recirculatory flow, namely marangoni convection, in the weld pool. Figure 3.14 shows schematically the model of marangoni convection driven by the temperature coefficient of the surface tension (Ohji *et al.*, 1990). Heiple and Roper (1981, 1982) suggested that an outward fluid flow with a wide and shallow weld was caused by a normal negative temperature coefficient of surface tension, whereas an inward fluid flow and resultant narrow and deep weld was caused by a positive temperature coefficient (see Fig. 3.14).

Recently, Tanaka *et al.* (2003) proposed a numerical model of the weld pool taking account of the close interaction between the arc plasma and the weld pool. The time-dependent development of the weld penetration was predicted at a current of 150 A in TIG welding of stainless steels containing low sulphur (40 ppm) and high sulphur (220 ppm). It was shown that calculated convective flow in the weld pool of an argon-shielded TIG process was dominated by the drag force of the cathode jet and the marangoni force. The other two driving forces, namely, the buoyancy force and the electromagnetic force, were significantly less important. Tanaka *et al.* (2003) also concluded that change in the direction of recirculatory flow in the weld pool led to dramatically different weld penetration geometry.

Gas tungsten arc welding

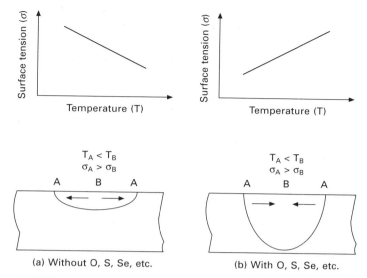

3.14 Schematic illustration of model of marangoni convection driven by temperature gradient of surface tension (Ohji *et al.*, 1990).

3.3.5 Current understanding of process mechanism

From Sections 3.3.3 and 3.3.4, the most plausible mechanism at present is proposed, as follows.

In the case of the conventional TIG process an outward fluid flow in the weld pool is caused by the drag force of the cathode jet and the marangoni force associated with a normal negative temperature coefficient of surface tension. This causes the heat input from the arc to transfer from the centre to the edge on the weld pool surface. This heat transfer leads to a shallow gradient in surface temperature across the weld pool. It also leads to the metal plasma distribution being expanded widely across the whole weld pool surface, owing to the much lower ionisation potential of the metal compared with that of the shielding gas. As a result, a diffuse anode root is formed. Strong convective flow outward at the surface of the weld pool leads to a shallower weld than heat transfer by conduction alone.

In the A-TIG process, the temperature coefficient of surface tension changes from negative to positive due to the surface active elements, such as oxygen, from the decomposition of the flux. The marangoni force associated with a positive temperature coefficient of surface tension is larger than the drag force of the cathode jet, and causes inward fluid flow. As a result of this inward flow, the heat input from the arc should transfer directly from the surface to the bottom of the weld pool. This heat transfer causes a steep gradient in the surface temperature of the weld pool, which also leads to the metal plasma distribution being localised at the centre on the weld pool surface. Consequently, a constricted anode root is formed and the anode spot appears to be located at the centre of the weld pool surface. Furthermore, the constricted anode root should lead to higher current density at the anode, which should also promote the inward recirculatory flow driven by the electromagnetic force. Thus the multiplication effect of the electromagnetic force and the marangoni force appears to cause strong inward recirculatory flow in the weld pool. Strong inward convective flow of the weld pool leads to a deeper weld than for heat transfer by conduction alone.

This proposed mechanism suggests that the deep weld penetration can be achieved by the activating flux even if the welding process is changed from TIG to alternative processes such as plasma, laser and electron beam. In fact Howse and Lucas (2000) have reported that the flux equally increased the depth of the weld penetration for both the plasma process and the laser process, although it did not in the case of the electron beam process. The electron beam did not show major increases in penetration as a result of the activating flux although the beam power density was modified to simulate that of a typical TIG arc (Howse and Lucas, 2000). However, Ohji *et al.* (1991) reported that the deep weld penetrations were achieved independently of sulphur contents (20 ppm, 60 ppm and 90 ppm) of the same type 304 stainless steels as enough defocused electron beam was employed for welding. In the electron beam process, only the marangoni force affects the convective flow in the weld pool because both the drag force of the cathode jet and the electromagnetic force can be neglected (Fujii *et al.* 2001). However, the marangoni force is strongly dependent not only on the temperature coefficient of surface tension but also on the sulphur (or oxygen) concentration coefficient of surface tension (Winkler *et al.*, 2000). Ohji *et al.* (1991) suggested that the latter coefficient of surface tension was very important for understanding the phenomena of the weld penetration in the electron beam process because the evaporation rate from the weld pool was much higher than that in TIG. This is due to the vacuum environment used in the electron beam process. The convective flow caused by the gradient of surface tension in liquids was first reported by James Thomson (Scriven and Sternling, 1960). Thomson (1855) provided the first correct explanation of the spreading of an alcohol drop on a water surface, the well-known 'tears of wine', and related phenomena. These phenomena are convective flows caused by changes in surface tension driven by the evaporation of the alcohol and the existence of an alcohol concentration coefficient of surface tension in the wine.

3.3.6 Examples of applications

Typical applications of A-TIG are precision welds in relatively thick (3–12 mm) material where advantage can be taken of a single pass. The A-TIG should find particular application in the orbital welding of tubes (Lucas and Howse, 1996). The tube can be welded in a single pass with a simple square butt joint, while three or more passes would be required with conventional TIG. The A-TIG process is also suitable for thinner wall material because the welds can be made at higher speeds and lower heat inputs than welds with conventional TIG (Lucas and Howse, 1996). For example, A-TIG can make a weld joint of 2 mm thickness in AISI 304 stainless steel at 800 mm/min welding speed, which is double that of conventional TIG, while the reduced heat input obviously results in less distortion (Okazaki and Okaniwa, 2002). Disadvantages of using A-TIG are the rougher surface appearance of the weld bead and the need to clean it after welding (Lucas, 2000). The as-welded surface is significantly less smooth than that produced with the conventional TIG, because there is significant slag residue on the surface of the weld produced with the A-TIG process. It often requires rigorous wire brushing to remove it (Lucas, 2000).

Some typical applications of the A-TIG process, such as in nuclear reactor components, car wheel rims, steel bottles and pressure vessels have been reported (Lucas and Howse, 1996). AISI type 316 stainless steel tube, 70 mm diameter and 5 mm wall thickness, was reported to be welded in a single pass with a simple square butt joint without filler wire by using the activating flux under the conditions of pulse current 150 A, background current 30 A, arc voltage 9.5 V and welding speed 60 mm/min (Lucas, 2000). Full weld penetration was achieved independently of the orbital positions.

In another report (Kamo et al., 2000) AISI type 304 stainless steel tube, 60.5 mm diameter and 8.7 mm wall thickness, with 4 mm root thickness and narrow-gap grove was reported to be welded using the activating flux. The welding conditions were welding current 100–130 A, arc voltage 10–11 V and welding speed 80–100 mm/min. The A-TIG required only three layers and three passes to complete the joint whereas conventional TIG required seven layers and seven passes. There were no defects such as insufficient fusion or cracking at any position, including the vertical-up and vertical-down positions, using the A-TIG process. The integrity of the weld joint produced by the A-TIG process was confirmed by both mechanical and metallographic tests (Kamo et al., 2000).

The A-TIG was also used for repairing cracks in metal at a nuclear power plant (Takahashi et al., 2002, Tsuboi et al., 2002). Type INCONEL600 nickel-based alloy was reported to be welded using an activating flux cored wire as a filler wire while welding underwater at double atmospheric pressure. The integrity of the weld joint and a deep weld penetration about 4.5 mm were achieved in this process.

Sire and Marya (2001) have proposed a new technique, called the FBTIG (flux bounded TIG) process, for producing a deep weld penetration in aluminium alloys. In this process, silica (SiO_2) flux was used to restrict the arc current to a narrow channel to enhance the weld penetration depth. The silica flux was pasted on the aluminium alloy surface, leaving a flux gap around the joint, whereas a fully flux coverage of the joint area was always maintained in the A-TIG process. The current was restricted to the gap due to the high electric resistance of the silica. It was possible to take full weld penetration of type 5086 aluminium alloy, 6 mm thickness, at 175 A and 150 mm/min with a flux gap of 4 mm.

3.4 The keyhole GTAW process

3.4.1 Introduction

Normally GTA welding of plates of more than a few millimetres in thickness calls for careful edge preparations and multiple passes. One route to achieving deeper penetration has been through the use of active fluxes, as described in the preceding section. This A-TIG process has advantages of simplicity and application across a very broad range of materials. An older and more direct approach has been to weld with much higher currents, as in the 'high current' GTAW process (Liptak 1965; Adonyi-Bucurdiu, 1989; Adonyi *et al.*, 1992). With this approach the increased arc forces push aside the liquid weld metal, allowing the arc to access regions well below the plate surface (Fig. 3.15). In practice, however, the degree of penetration is difficult to control, and with rising current the pool becomes increasingly unstable with respect to

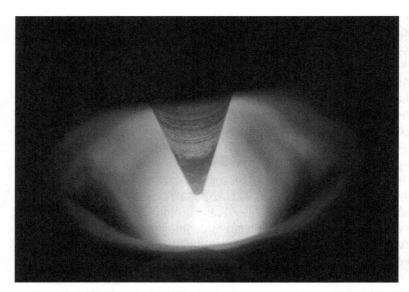

3.15 Close-up of a high current gas tungsten arc displacing the weld pool through the action of arc forces.

Gas tungsten arc welding

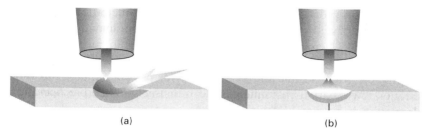

3.16 Schematic of conventional melt-in mode gas tungsten arc welding: side (a) and front (b) views.

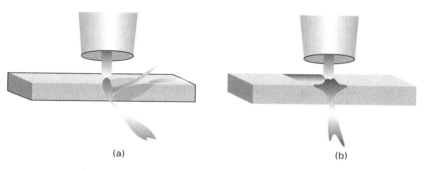

3.17 Schematic of keyhole-mode gas tungsten arc welding: side (a) and front (b) views.

fluctuations in arc pressure over its surface. Hollow-tipped tungsten electrodes have been developed as one means of reducing arc pressure and improving stability (Yamauchi *et al.*, 1981).

Another variant, 'keyhole GTAW' is now attracting industrial attention. Keyhole GTAW differs from other modes in forcing an opening all the way through the joint. Despite this it still completes the weld without the need of a backing bar. This difference is illustrated schematically in Fig. 3.16 and 3.17. The novelty of the process arises both through the peculiar choice of operating conditions and in the use of a torch designed to deliver high axial arc pressures under very stable and reproducible conditions. The process can be implemented using 'off-the-shelf' GTAW power sources of suitable rating (a 600 A supply would be suitable for most applications). Enhancements designed to pinch or otherwise constrict the arc are not used.

The process was first introduced to commercial applications in Australia in the late 1990s, but is now finding use around the globe. It is applicable to a wide range of lower conductivity metals and alloys, (e.g. steels, stainless steels, titanium and nickel alloys). Applications requiring long, full-penetration butt welds such as spiral and seam welded pipe would seem to be ideal candidates for the process. In its present state of development it is not suited to highly conductive metals such as copper and aluminium.

3.4.2 Process performance

Keyhole GTAW is easy to implement and can be used within broad operating windows. Its primary attraction is that it is a fast single pass process, providing, for example, full penetration of stainless steel plates from 3 mm to about 12 mm thick – and to about 16 mm for titanium alloys. This is achieved using only minimal edge preparation and filler material because joints are presented in closed square-butt configuration. Such performance represents a significant advantage over GMAW and conventional GTAW for many applications. Similar performance may be obtained with plasma arc welding, but implementation costs are greater and process operation is more complex.

Control over the process is exercised through variations to the electrode geometry, voltage, current, travel speed and shielding gas composition. In general use, however, most parameters are fixed, with subsequent variation of only travel speed and current being sufficient to access most of the operating window. Furthermore, keyhole operation is readily confirmed through observation of the efflux plasma emerging from the root face and this has been used as a simple but effective control strategy.

Keyhole gas tungsten arc welds are not unlike plasma keyhole welds in appearance. Typically, the width of the weld crown is slightly greater than the thickness of the plate, providing an aspect ratio (depth to width) between 0.5 and 1.0. The root bead width tends to be between 2 and 4 mm. When sectioned, the fusion boundary is found to be slightly concave rather than straight. The fusion zones also display often pronounced caps or 'nail-heads' giving the impression of two or more passes having been applied. In fact the nail-head results from some additional fusion that occurs in the tail of the weld pool due to the accumulation of superheated weld metal. This can be deduced through inspection of the crater of an abruptly terminated keyhole weld. There it can be seen that the width of the weld increases in the trailing portion of the weld pool.

3.4.3 Conventional view of keyholes in welding

Keyhole welding is usually identified with laser and electron beam processes. These two processes are known for their deep narrow keyholes, often with aspect ratios exceeding 10:1. Such keyhole cavities are approximately cylindrical and so have a strong tendency to collapse under the pressure due to surface tension (γ/r for a cylindrical geometry) and the head of liquid metal ($\rho g h$). (In these expressions γ is surface tension, r the radius of the keyhole channel, ρ the liquid metal density, g the acceleration due to gravity and h the depth). The accepted view is that these keyholes are maintained by an increased pressure in the cavity generated by the recoil from ablating material (Andrews and Atthey, 1976; Lancaster, 1986; Matsunawa et al., 1998). Ablation and its associated recoil pressure only becomes significant at extremely high power densities – exceeding 10^9 W/m². Even so, the narrowness of the laser and electron beam channels calls for power densities an order of magnitude higher (i.e. 10^{10} W/m²). By way of comparison, an oxy-acetylene flame can achieve a power density of about 1×10^7 W/m² (Jarvis, 2001) and GMA and GTA arcs a little over 1×10^8 W/m² (Lancaster, 1986).

The plasma arc process (PAW) has been regarded as the only arc welding process typically operated in a keyhole mode and this has often been cited as its primary advantage over GTAW (*ASM Handbook*, Vol 6; Halmoy, 1994). Nevertheless, there have been descriptions of successful keyhole welding using modified GTAW equipment, for example in the dual-gas GTAW process (Norrish, 1992). In this case, however, the shielding gas arrangement has similarities to that of the PAW process in that it produces a strong 'thermal pinch' to increase the power density of the arc. PAW processes can achieve power densities of about 3×10^9 W/m² and produce wider, lower aspect ratio keyholes than the laser and electron beam processes. It should be noted, however, that a significant portion of the pressure needed to stabilise PAW keyholes results from the mechanical impact of the plasma jet (Lancaster, 1986).

PAW keyholes must be 'open' to allow the venting of the arc gases. This means that the keyhole must extend all the way through to the root face. The plasma escaping from the bottom of the keyhole is referred to as the 'efflux' plasma.

3.4.4 The GTAW keyhole

With the possible exception cited above, GTAW is believed to be incapable of delivering the power densities required to generate appreciable recoil pressures (Lancaster, 1986). Keyhole GTAW has not received significant scrutiny in the literature as yet, and therefore the dominating mechanism has not been widely debated. Nevertheless, it is theorised that these low aspect ratio keyholes (aspect ratios are often less than one) are stabilised by surface tension. This being the case, it can be argued that arc pressure alone is sufficient to establish and maintain the keyhole (Jarvis, 2001). The proposed model predicts that if the keyhole surfaces could be accurately reconstructed they would be found to be closely related to minimal surfaces, familiar in such phenomena as soap films.

To begin, consider the link between surface tension and pressure. The pressure, P, across a liquid surface due to the surface tension is related to the surface curvature and can be written in the following form (Laplace's equation):

$$P = \gamma\left(\frac{1}{r_1} + \frac{1}{r_2}\right) \quad [3.8]$$

In this expression r_1 and r_2 are the principal radii of curvature.

Principal axes may be chosen arbitrarily provided only that they are orthogonal and tangent to the surface at the chosen point. This is due to a result from geometry that states that 'at any point on any surface the sum of the reciprocals of the radii of curvature in any two mutually perpendicular sections is constant' (Grimsehl, 1947). Furthermore, the sum of the two reciprocals is called the total curvature of the surface at that point. If this result is applied to Laplace's equation (above) then, 'the normal pressure due to surface tension at any point on a surface is equal to the product of surface tension and total curvature' (Grimsehl, 1947). That is, if K signifies total curvature then

$$K = \frac{1}{r_1} + \frac{1}{r_2} \quad [3.9]$$

and

$$P = \gamma K \quad [3.10]$$

In general the two radii need not be of similar value, or even of the same sign. For illustration, if the surface is cylindrical, one radius will be infinite and can be ignored, (and $K = 1/r$) as is often the case in laser and electron beam keyholes. Alternatively if the surface is spherical, as in a bubble, both radii are equal (and $K = 2/r$). Both cylindrical and spherical surfaces will collapse unless there is a counteracting pressure because in these cases the curvature can never be zero. However, it is also possible for a surface to be concave along one axis and convex along another, that is, r_1 and r_2 have opposing signs. The resultant pressure change across the surface then may be positive, negative or zero. If the net pressure change is zero, the surface will be stable. Furthermore, if surface tension is the only force acting then the surface will be a 'minimal surface'. As can be anticipated from Fig. 3.18, the stability of this type of surface is dependent on its aspect ratio.

If one is dealing with a free surface then there is no net force acting anywhere over the surface, i.e.,

$$\gamma K + \rho g z + P_{arc} + P_{inertial} + \ldots = 0 \quad [3.11]$$

where z is the distance below the pool surface, P_{arc} is the arc pressure, $P_{inertial}$ is a pressure that might be anticipated in a moving weld pool and so on. (Note that in the application of this equation care must be exercised in determining the signs of the various terms).

Since such free surfaces have not been widely discussed in relation to welding it is useful to look a little more closely at the details.

Gas tungsten arc welding 69

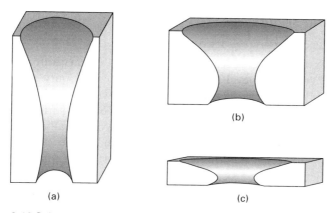

3.18 Schematic illustration of how the two radii of curvature identified in the text vary with aspect ratio. Only (b) is expected to be stable.

Potential for surface tension stabilised keyholes

To illustrate the problem, consider the simplified system of a stationary, axisymmetric arc and pool. Let the surface be described in cylindrical coordinates. Put $z = 0$ at the surface of the plate, with z decreasing with depth. Further, let the pool surface make an angle θ with the radial vector (See Fig. 3.19).

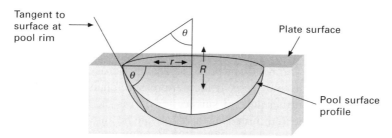

3.19 The identification of the variables used in estimating the maximum force due to surface tension.

Now the vertical force due to surface tension acting on the surface at a radius r is

$$F = \gamma 2\pi r \sin \theta \qquad [3.12]$$

In order that there is no net force acting on the surface, except at the boundaries, the change in F over the annulus between the two radii, r and $r + dr$, must equal the vertical component of force due to all other pressure terms ($P(r)$) over the same area. Thus:

$$2\pi\gamma \, d(r \sin \theta) = 2\pi r \, P(r) dr \qquad [3.13]$$

This gives the differential equation of the surface

$$\frac{d}{dr}(r \sin \theta) = \frac{rP(r)}{\gamma} \qquad [3.14]$$

It is now convenient to express $\sin \theta$ in terms of z' ($= dz/dr$):

$$\sin \theta = \frac{z'}{\sqrt{1 + z'^2}} \qquad [3.15]$$

to give

$$\frac{d}{dr}\left(\frac{rz'}{\sqrt{1 + z'^2}}\right) = \frac{rP(r)}{\gamma} \qquad [3.16]$$

An important mathematical case arises when $P(r)$ is set to zero. (In welding this would be related to the improbable situation in which the arc and hydrostatic pressures are everywhere in balance).

Setting $P(r)$ to zero and integrating once gives

$$\frac{rz'}{\sqrt{1 + z'^2}} = c \qquad [3.17]$$

Which on solving for z' gives

$$z' = \frac{c}{\sqrt{(r^2 - c^2)}} \qquad [3.18]$$

If $c = 0$, the solution is $z =$ constant. This corresponds to a flat weld pool and is of little interest. However, if $c \neq 0$ then the solution is

$$z = c \cosh^{-1}\left(\frac{r}{c}\right) + d \qquad [3.19]$$

where d is another constant of integration. Rearranging to make r the dependent variable:

$$r = c \cosh\left(\frac{z - d}{c}\right) \qquad [3.20]$$

This is the equation of a catenoid (see Fig. 3.20). This solution is particularly important because it can be regarded as a 'surface tension stabilised keyhole' and therefore supports the notion of keyhole solutions for weld pool surfaces. In practical terms it is supposed that a depressed pool surface becomes stabilised when it is pushed deeply enough to attach to the root face and rupture, creating an opening from the front to root faces.

If equation [3.16] is rearranged to represent pure pressure terms on each side

Gas tungsten arc welding

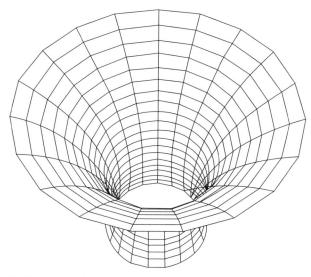

3.20 A catenoid. This is an example of a minimal surface and as such embodies the principles of the GTAW keyhole stability.

$$\frac{\gamma}{r}\frac{d}{dr}\left(\frac{rz'}{\sqrt{1+z'^2}}\right) = P(r) \qquad [3.21]$$

Then one finds from [3.10] that the surface curvature (K) of an axi-symmetric surface can be expressed as

$$K = \frac{1}{r}\frac{d}{dr}\left(\frac{rz'}{\sqrt{1+z'^2}}\right) \qquad [3.22]$$

Alternatively, the curvature can be determined using the more general methods of the calculus of variations (Jarvis, 2001). This approach reproduces the above expression, or, if z is made the independent variable,

$$K = \frac{1}{r}\left(\frac{d}{dz}\frac{rr'}{\sqrt{1+r'^2}} - \sqrt{1+r'^2}\right)$$

$$= \frac{rr'' - r'^2 - 1}{r(1+r'^2)^{3/2}} \qquad [3.23]$$

The work that has been done to date has clearly established that keyhole solutions do exist for much more realistic approximations to certain GTA welding situations (Jarvis, 2001). However, considerably more effort is required to extend these results to situations involving complexities such as reduced symmetry, metal flow and arc–weld pool interactions.

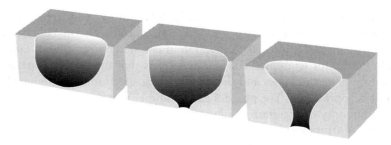

3.21 Schematic illustrating the transitions between deep cavity melt-in mode and keyhole mode. The former is held open by arc forces, while the more stable keyhole is held open by arc forces and surface tension. The transitional state (centre) is presumed unstable.

An interesting aspect of these propositions is that keyhole generation involves the transition through fundamentally different surface geometries. When the weld pool is significantly depressed, but not in a keyhole mode, surface tension acts to resist the deformation and the pool can be quite unstable. Once the surface is ruptured, forming an opening between the top and root surfaces of the plate, surface tension can drive the pool to a very stable keyhole geometry (see Fig. 3.21). One result is that the process may exhibit some hysteresis. This means that a keyhole may not form until the welding current is raised to a threshold value, but once formed it may remain open even if the current is reduced below the threshold.

An extension of the keyhole concept may offer an explanation for certain types of porosity, including tunnel porosity, that are a common defect in high current GTA welding (Jarvis, 2001). Keyhole surfaces may also generate potentially strong driving forces for (and coupling with) metal flow, as argued in Section 3.2.3. In light of these various possibilities it would appear that the study of free surface behaviour could be a fruitful area for welding science.

3.4.5 Formation of a root bead

The stability of the root bead can also be attributed to surface tension. For example, it is noted that the radius of curvature across the root bead (r_w) must be greater than half the width of the root bead, w (i.e. $r_w \geq w/z$). Typically, w/z is between 1 and 2 mm. On the other hand, the radius of curvature along the solidifying bead (r_a) can be very large at high welding speeds, implying that the maximum pressure that can be sustained at high welding speeds is approximately $2\gamma/w$. This pressure can be balanced against that due to the head of liquid metal, $\rho g h$. Inserting realistic values for w and γ indicates that those keyhole welds in AISI 304 stainless steel thicker than about 7 mm will not be supported at high welding speeds. However, if the welding speed, and hence r_a, is reduced, stability might be restored. Of course there is a limit to the reduction in r_a – it certainly cannot take values below $w/2$, for example. The implication is that the process operates to a limiting plate thickness and that the welding speed must be reduced as this limit is approached. The process can handle 12 mm thick austenitic stainless steel but only over a narrow range in welding speed that certainly does not exceed 400 mm/minute.

The surface tension also appears to limit the minimum thickness of plate that can be welded. In this case the limit is established by geometric considerations associated with the width of the weld pool and is not directly dependent on material properties. Failure results in cutting of the plate. In practice, the process is very difficult to operate with materials less than 3 mm thick. However, such thicknesses are readily accommodated with more conventional welding modes, leaving little incentive to develop practical solutions.

3.4.6 Arc–keyhole relations

As outlined earlier (Section 3.2.2) high current welding arcs exert noticeable forces on the weld pools and tend to push the molten metal aside. An expression relating the total arc force to the anode and cathode radii (r_a and r_e) developed by Converti (1981) is

$$F = \frac{\mu I^2}{8\pi}\left(1 + 2 \ln \frac{r_a}{r_e}\right) \qquad [3.24]$$

The ratio r_a/r_e is sometimes referred to as the arc expansion ratio.

The principal consideration in generating the conditions necessary for keyhole GTAW is the production of a high peak arc pressure and hence the minimisation of the cathode emission radius r_e.

In keyhole GTAW the cathode emission region is confined to the tip of the tungsten electrode. The electrons emitted from this region maintain the current in the arc plasma. There are various mechanisms by which electrons can pass from an electrode into a surrounding gas or plasma. In this case the mechanism is thermal; the electrode tip is so hot that electrons can 'evaporate' into the surrounding gas from where the arc voltage drags them through the plasma to the weld pool. Consequently, the area of this region is sensitive to the heat flow within the electrode, and this in turn affects the current at which a keyhole can form (this is referred to as the threshold current). Thus changing the electrode stick-out, taper, diameter or composition are all means of altering the point of transition to keyhole mode. Choosing large diameter electrodes is one of several simple strategies for reducing the area of emission and therefore the threshold current.

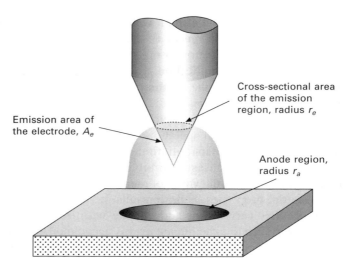

3.22 Schematic diagram illustrating the parameters used to examine the effects of electrode geometry on threshold current. The tip has an included angle of $\theta°$.

The liberation of electrons from the tip requires a great deal of energy, and so the process cools the emission region. As a result the region tends towards a constant temperature, and its area is strongly correlated with the arc current. However, the peak arc pressure is dependent on the cross-sectional radius of this region and not on its area. The implication from this is that the peak arc pressure can be increased significantly by reducing the included angle of the electrode tip. This is because the cross-sectional radius of the emission region, r_e is a function of both the emission area, A_e and the included angle, θ (see Fig. 3.22):

$$r_e = \sqrt{\frac{A_e \sin\left(\frac{\theta}{2}\right)}{\pi}} \qquad [3.25]$$

In practice the included angle is usually kept between 45° and 60°.

Helium-rich arcs are not desirable for keyhole welding because, apparently, the high viscosity of helium dampens the action of the arc core. However, they are very effective in transferring thermal energy to the weld pool and this can be advantageous. Fortunately there are means of obtaining high conductivity without reducing the arc pressure. In particular, diatomic gases absorb substantial amounts of energy in dissociation and this provides a highly efficient energy transport mechanism known as 'reactive thermal conductivity'. The two most likely candidate gases are hydrogen and nitrogen.

These can be added to argon to give significantly better keyholing potential than argon alone. Hydrogen additions of up to about 10% can be used with austenitic stainless steels, while similar concentrations of nitrogen have been used with duplex stainless steels. Naturally, such additions cannot be used indiscriminately, as they can have seriously detrimental effects on many metals and alloys.

The final arc parameter of interest is arc length. Arc length has a significant effect on keyhole behaviour when operation is near the limits of the process envelope. Also, keyholes tend not to form when the electrode tip is submerged. In one set of trials the electrode was incrementally raised from a submerged position. At first a keyhole could not be formed. Keyholing only became possible when the tip was raised to be approximately level with the plate surface, but the required current was high. However, on continued increase in arc length the threshold current exhibited a rapid transition to a significantly lower value. This lower level then remained constant on continued raising of the electrode (see Fig. 3.23). The implication was that operation with too short an arc length can give inconsistent performance, with the threshold current either becoming random between the two levels, or the process failing completely. At the other extreme of arc length the weld pool may also broaden, this time due to expansion of the arc. Thinner (e.g. 6mm) plate does not appear to be sensitive to this but thicker sections have shown quite well-defined upper as well as lower limits to arc length, and by association, arc voltage.

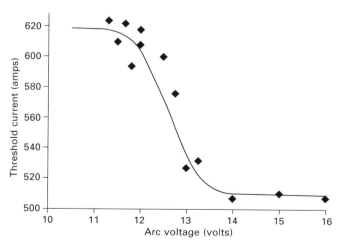

3.23 Threshold current for keyhole mode as a function of voltage for 5.1mm SAF 2205 (travel speed 300mm/min, 3.2mm electrode).

3.4.7 Summary

Keyhole mode gas tungsten arc welding is a new process variant with considerable potential. It is able to capture some of the best features of the GTAW process (cleanliness, controllability and versatility), while adding much sought-after productivity gains. Laboratory and industrial experience suggests that this is an attractive option for applications involving automated flat position welding of materials above 3 mm thickness.

However, the process has not been widely studied and although it appears that there is a basic understanding of the mechanics and operational details, there is scope for considerably more analysis and development. For example, issues relating to highly distorted free surfaces are central to understanding the keyhole, but are not well-understood from a welding perspective. Given the perceived potential for this process it is hoped that it will receive more attention in the future.

3.5 Future trends

There are a number of misconceptions and genuine limitations relating to GTAW and these must be addressed if the process is to retain its relevance in the future. The 'basic' GTAW process has been hampered by its low penetration and consequent poor productivity. As a production tool it tends to be used when quality or other overriding issues demand it. This chapter has argued that the process has much more to offer and has illustrated this with the detailed description of two of its many variants. It is suggested that this realisation that GTAW has 'more to offer' will be increasingly appreciated, particularly as fabrication operations become more integrated and mechanised. One of the historical impediments to the seamless integration of welding into production lines has been poor joint fitup and the consequent need for a degree of adaptability that was only available with manual intervention. This impediment is rapidly being removed as component tolerances improve, welding processes become more tolerant, and control systems are made more intelligent and responsive. This trend will suit the lower deposition welding processes such as GTAW and should renew the search for innovative ways of exploiting this very elegant process.

Many of the changes to GTAW over recent times have been forecast correctly to be in the area of the equipment used to implement the process (Lucas, 1990; Muncaster, 1991). This area covers power sources, control systems, monitoring, viewing and data acquisition (Muncaster, 1991). This trend is expected to continue in the future, with the increasing availability of significant computational power driving the process in the direction of greater adaptability and user friendliness. Occupational health and safety as well as environmental issues are also becoming more important and concerns about electromagnetic radiation and its potential to interfere with computerised equipment, metal fume and overall power requirements, will all lead to further changes in equipment and practices.

However, the opportunity for new variants is expected to continue and to produce some very productive processes. One example of this is the recent research into hybrid processes and particularly laser plus GTAW (Dilthey and Keller, 2001; Ishide et al., 2002). Hybrid welding refers to a situation where two processes (in this case, laser welding and GTAW) are coupled together to act at a single point. The coupling between the laser beam and the gas tungsten arc produces a number of synergistic effects that enhance the best features of each process. For example, the laser not only provides deep penetration but also stabilises the anode spot of the arc. As one result the gas tungsten arc then can be operated in the more efficient DCEN mode, even when welding aluminium. At the same time the arc broadens the weld pool at the plate surface, improves the laser to material coupling, and relaxes the very high joint tolerances required for laser welding. It also provides additional heat input and an improved weld profile with reduced notch angles. In one set of trials on a 2 mm aluminium 3% magnesium alloy, Dilthey and Keller (2001) reported an increase in welding speed from 5 m/min for the laser, to 8 m/min with the hybrid process. The GTAW operated alone could only be operated in the ac mode at 2 m/min.

Another innovation, in GTAW is the newly reported guided GTAW or GGTAW process (Zhang et al., 2003). In this variant the main arc is established between a short, hollow tungsten electrode and the workpiece. However, a separately powered electrode positioned above the main electrode provides a lower current 'pilot arc'. This arc is constricted in passing through the hollow main electrode. The result is two concentric arcs, the inner of which has a high energy density and is relatively stiff. The inner arc has the effect of stiffening or 'guiding' the main arc, hence the name of the process. This process is anticipated to have some advantages over both GTAW and plasma arc.

In summary, GTAW is a particularly elegant welding process because of its apparent simplicity and appeal to fundamental physical principles. It is also becoming far more productive and versatile than popular images of the process suggest. The likely scenario is that this process will continue to be developed in new and imaginative ways for many years to come.

3.6   References

Adonyi Y., Richardson R.W. and Baeslack III W.A. (1992), 'Investigation of arc force effects in subsurface GTA welding', *Welding Journal*, **71**(9) 321s–30s

Adonyi-Bucurdiu I. (1989), *A Study of Arc Force Effects During Submerged Gas Tungsten-arc Welding*, PhD Dissertation, The Ohio State University Ohio

Anderson P.C.J. and Wiktorowicz R. (1996), 'Improving productivity with A-TIG welding', *Welding and Metal Fabrication*, (3/12) 108–9

Andrews J.G. and Atthey D.R. (1976), 'Hydrodynamic limit to penetration of a material by a high-power beam', *Journal of Physics D: Applied Physics*, **9** 2181–94.

Block-bolten A. and Eagar T.W. (1984), 'Metal vaporization from weld pools', *Metallurgical Trans. B*, **15B**(9) 461–469

Converti J. (1981), *Plasma Jets in Welding Arcs*, PhD Thesis, Mechanical Engineering, MIT, Cambridge, MA

Dilthey U. and Keller H. (2001), 'Laser arc hybrid welding', *Proc. 7th Int. Welding Symposium*, Kobe, Japan Welding Soc.

Erokhin A.A. (1979), 'Force exerted by the arc on the metal being melted', *Avtom. Svarka*, **7** 21–6

Ferjutz K., Davis J.R. and Wheaton N.D. (eds) (1994), ASM Handbook, Volume 6, *Welding Brazing and Soldering*. ASM International, 195–9

Fujii H., Sogabe N., Kamai M. and Nogi K. (2001), 'Effects of surface tension and gravity on convection in molten pool during electron beam welding', *Proc. 7th Int. Welding Symposium*, Kobe, Japan Welding Soc., 131–6

Grimsehl E. (1947), *A Textbook of Physics, Vol 1, Mechanics*, London, Blackie & Son.

Gurevich S.M. and Zamkov V.N. (1966), 'Welding titanium with a non-consumable electrode using fluxes', *Automat. Welding*, **12** 13–16

Gurevich S.M., Zamkov V.N. and Kushnirenko N.A. (1965), 'Improving the penetration of titanium alloys when they are welded by argon tungsten arc process', *Automat. Welding*, **9** 1–4

Halmoy E. (1994), 'New applications of plasma keyhole welding', *Welding in the World*, **34**, 285–91

Heiple C.R. and Roper J.R. (1981), 'Effects of selenium on GTA fusion zone geometry', *Welding Journal*, 60 143s–5s

Heiple C.R. and Roper J.R. (1982), 'Mechanism for minor element on GTA fusion zone geometry', *Welding Journal*, **61** 97s–102s

Howse D. and Lucas W. (2000), 'Investigation into arc constriction by active fluxes for tungsten inert gas welding', *Sci. and Tech. Welding and Joining*, **5**(3) 189–93

Ishide T., Tsubbota S., Watanabe M. and Ueshiro K. (2002), *Development of YAG Laser and Arc Hybrid Welding Method*, Int. Inst. Welding Document Doc No. XII-1705–02

Jackson C.E. (1960), 'The science of arc welding', *Welding Journal*, **39**(4), 129s–140s and **39**(6), 225s–30s

Jarvis B.L. (2001), *Keyhole Gas Tungsten Arc Welding: a novel process variant*, PhD Thesis, Mechanical Engineering, University of Wollongong, Wollongong

Kamo K., Nagura Y., Toyoda M., Matsubayashi K. and Miyake K. (2000), 'Application of GTA welding with activating flux', *Intermediate Meeting of Comm. XII of International Institute Welding*, Ohio, Int. Inst. Welding, Doc. No. XII-1616–00

Katayama S., Mizutani M. and Matsunawa A. (2001), 'Liquid flow inside molten pool during TIG welding and formation mechanism of bubble and porosity', *Proc. 7th Int. Welding Symp.*, Kobe, Japan Welding Soc., 125–30

Lancaster J.F. (1986), *The Physics of Welding*, 2nd ed., IIW publication, Oxford and New York etc., Pergamon Press

Liptak J.A. (1965), 'Gas tungsten arc welding heavy aluminium plate', *Welding Journal*, **44**(6) 276s–81s

Lorrain P. and Corson D. (1970), *Electromagnetic fields and waves*, 2nd ed., San Francisco, W.H. Freeman and Co.

Lowke J.J., Kovitya P. and Schmidt H.P. (1992), 'Theory of free-burning arc columns including the influence of the cathode', *Journal of Physics D: Applied Physics*, **25**(11) 1600–6

Lu S.-P., Fujii H., Sugiyama H., Tanaka M. and Nogi K. (2002), 'Weld penetration and marangoni convection with oxide fluxes in GTA welding', *Materials Trans. of Japan Institute of Metals*, **43**(11) 2926–31

Lucas W. (1990), *TIG and Plasma Welding*, Cambridge, UK, Woodhead Publishing Ltd

Lucas W. (2000), 'Activating flux – improving the performance of the TIG process', *Welding and Metal Fabrication*, (2/12) 7–10

Lucas W. and Howse D. (1996), 'Activating flux – increasing the performance and productivity of the TIG and plasma process', *Welding and Metal Fabrication*, (1/12) 11–17

Lucas W., Howse D., Savitsky M.M. and Kovalenko I.V. (1996), 'A-TIG flux for increasing the performance and productivity of welding processes', *Proc. 49th International Institute of Welding Annual Assembly*, Budapest, Hungary, Int. Inst. Welding, Doc. No. XII-1448–96

Makara A.M., Savitskii M.M. and Kushnirenko B.N. et al. (1977), 'The effect of refining on the penetration of metal in arc welding', *Automat. Welding*, **9**, 7–10

Matsuda F., Ushio M. and Sadek A. (1990), 'Development of GTA electrode materials'. *The 5th International Symposium of the Japanese Welding Society*. Tokyo, April 1990

Matsunawa A. (1992), 'Modelling of heat and fluid flow in arc welding', Keynote address, *International Trends in Welding Conference*, Gattlingburg, 1992

Matsunawa A., Kim J-D., Seto N., Mizutani M. and Katayama S. (1998), 'Dynamics of keyhole and molten pool in laser welding', *Laser Applications*, **10**(6) 247–54

Muncaster P. (1991), *Practical TIG (GTA) Welding*, Cambridge, UK, Abington Publishing

Norrish J. (1992), *Advanced Welding Processes*, Bristol, IOP Publishing Ltd

Ogino K., Nogi K. and Hosoi C. (1983), 'Surface tension of molten Fe–O–S Alloy', *Tetsu-to-Hagane (J. Iron Steel Inst. Japan)*, **69**(16) 1989–94

Ohji T., Miyake A., Tamura M., Inoue H. and Nishiguchi K. (1990), 'Minor element effect on weld penetration', *Q. J. Japan Welding Soc.*, **8**(1) 54–8

Ohji T., Inoue H. and Nishiguchi K. (1991), 'Metal flow in molten pool by defocused electron beam', *Q. J. Japan Welding Soc.*, **9**(4) 501–6

Okazaki T. and Okaniwa T. (2002), 'Application of active flux TIG welding', *J. Japan Welding Soc.*, **71**(2) 100–3

Ootsuki M., Tsuboi R., Takahashi H., Asai S., Taki K. and Makino Y. (2000), 'Study on high penetration welding using activated flux method' (1), *Preprints of the national meeting of Japan Welding Soc.*, **66** 240–1

Ostrovskii O.E., Kryukovskii V.N., Buk B.B. et al. (1977), 'The effect of activating fluxes on the penetration capability of the welding arc and the energy concentration in the anode spot', *Welding Production*, **3** 3–4

Papoular R. (1965), *Electrical Phenomena in Gases*, London, Iliffe Books

Savitskii M.M. (1979), 'The current density in the anode spot during the welding of standard and refined steels', *Automat. Welding*, **7** 17–20

Savitskii M.M. and Leskov G.I. (1980), 'The mechanism of electrically-negative elements on the penetrating power of an arc with a tungsten cathode', *Automat. Welding*, **9** 17–22

Scriven L.E. and Sternling C V. (1960), 'The marangoni effects', *Nature*, **187**(July) 186–8

Shaw Jr C.B. (1975), 'Diagnostic studies of the GTAW arc, Part 1' *Welding Journal*, **54**(2) 33s–44s

Simonik A.G., Petviashvili V.I. and Ivanov A.A. (1976), 'The effect of contraction of the arc discharge upon the introduction of electro-negative elements', *Welding Production*, **3** 49–51

Sire S. and Marya S. (2001), 'New perspectives in TIG welding of aluminum through flux application FBTIG process', *Proc. 7th Int. Welding Symp.*, Kobe, Japan Welding Soc., 113–18

Taimatsu H., Nogi K. and Ogino K.J. (1992), 'Surface tension of liquid Fe–O alloy', *High Temperature Soc. Japan*, **18** 14–19

Takahashi H., Asai S., Tsuboi R., Kobayashi M., Yasuda T., Ogawa T. and Takebayashi H. (2002), 'Study on underwater GTAW with active flux cored wire (1)', *Preprints of the National Meeting of Japan Welding Soc.*, **70** 24–5

Tanaka M. (2002), 'Effects of activating flux on weld penetration in TIG welding', *J. Japan Welding Soc.*, **71**(2) 95–9

Tanaka M., Shimizu T., Terasaki H., Ushio M., Koshi-ishi F. and Yang C.-L. (2000), 'Effects of activating flux on arc phenomena in gas tungsten arc welding', *Sci. and Tech. Welding and Joining*, **5**(6) 397–402

Tanaka M., Terasaki H., Ushio M. and Lowke J.J. (2003), 'Numerical study of a free-burning argon arc with anode melting', *Plasma Chem. and Plasma Process.*, **23**(3) 585–606

Thomson J. (1855), 'On certain curious motions observable at the surfaces of wine and other alcoholic liquors', *Philosophical Magazine*, **10** 330–3

Tsuboi R., Asai S., Taki K., Ogawa T., Yasuda T. and Takebayashi H. (2002), 'Study on underwater GTAW with active flux cored wire (2)', *Preprints of the National Meeting of Japan Welding Soc.*, **71** 264–5

Winkler C., Amberg G., Inoue H., Koseki T. and Fuji M. (2000), 'Effect of surfactant redistribution on weld pool shape during gas tungsten arc welding', *Sci. and Tech. Welding and Joining*, **5**(1) 8–20

Xiao Y.H. and den Ouden G. (1990), 'A study of GTA weld pool oscillation', *Welding Journal*, **69**(8) 289s–293s

Yamauchi N., Taka T. and Oh-I M. (1981), 'Development and application of high current TIG process (SHOLTA) welding process', *The Sumitomo Search*, **25** May 87–100

Zhang Y., Lu, W. and Liu Y. (2003), 'Guided arc enhances GTAW', *Welding Journal*, **12**, 40–5

Zhu P., Lowke J.J. and Morrow R. (1992), 'A unified theory of free burning arcs, cathode sheaths and cathodes', *Journal of Physics D: Applied Physics*, **25** 1221–30

4
Laser beam welding

V. MERCHANT, Consultant, Canada

4.1 Introduction: process principles

4.1.1 Brief history of lasers and laser beam welding

The first operating laser, built by Thomas Maiman in 1960, was a pulsed ruby laser producing millisecond long pulses with a low repetition rate, in the far red region of the spectrum. This laser used the intense light from flashlamps to excite the chromium atoms doped into a crystalline aluminum oxide rod; it is these chromium atoms that give ruby and synthetic ruby their distinctive colour. Typical rods may be 6 to 10 mm in diameter and 20 mm long. The excited chromium atoms radiate their excess energy as the red light, which is repeatedly reflected by carefully aligned mirrors at each end of the ruby rod, is passed multiple times through the rod and is amplified by the process of stimulated emission. Within a few years it was found that greater powers could be achieved by using, instead of ruby, a rod consisting of neodymium ions doped in otherwise very pure glass. Although the Nd:glass laser continues to be used where very high pulsed energies are required, it has been superseded in many applications by lasers built with rods consisting of neodymium atoms doped into a crystal of yttrium aluminium garnet, or Nd:YAG (Koechner, 1976).

The carbon dioxide laser, a gas laser with the potential to be scaled to high average powers, was invented by C.K.N. Patel in 1966. The simplest carbon dioxide lasers consist of a tube from which the air has been evacuated and replaced with a low pressure mixture of carbon dioxide, helium and nitrogen gases. Electrical current from a high voltage power supply or a radio frequency (RF) generator passes through the gas, exciting the carbon dioxide molecules. The mechanisms by which energy is transferred to the carbon dioxide molecule and optical output produced are discussed by Patel (1969) and by DeMaria (1976). High power carbon dioxide lasers require complex gas flow systems to circulate the gas excited by the electrical discharge through heat exchangers that extract the waste heat.

Early in the history of lasers, it was discovered that the laser beam output could heat, melt and vaporize metals. If the laser output was carefully controlled, the melting and subsequent solidification would result in welds between adjacent pieces of metal. Thus laser beam welding was born and announced almost simultaneously by three different suppliers of laser equipment who were seeking to expand their markets (Banas, 1972; Locke *et al.*, 1972; Bolin, 1976).

As of the early years of the twenty-first century, most laser beam welding is conducted by the output of either the carbon dioxide laser or the Nd:YAG laser. Both of these lasers, depending on the electrical excitation circuitry, can emit their output either continuously, as a single pulse, or as a repetitive series of pulses. Laser beam welding has been conducted with both continuous and pulsed lasers.

4.1.2 Laser output

A very large number of materials have been found to give laser output. The output of the light source known as a laser is very special and differs in many important aspects from the output of any non-laser source. The most important aspect of a laser source is its coherence; coherence implies a definite relationship between the output observed in different places and different times (on a microscopic scale) that results from the process of stimulated emission. In stimulated emission, one molecule with excess energy is stimulated to give up this energy when it is impinged by light of a particular wavelength. The light the excited molecule emits is in the same direction, the same polarization, the same phase and the same wavelength as the light that stimulated the emission. Since this light is reflected by the laser mirrors, all succeeding light that is emitted is in the same direction, the same polarization, the same phase and the same wavelength.

Of all the properties mentioned above, for laser welding the only relevant property is that the light emitted by the stimulated emission, or laser action, is in the same direction. By contrast, the light emitted by a fluorescent or incandescent bulb spreads out all over a room and is useful for a different purpose, that is, illuminating the room. The laser light is unidirectional and can be steered by a series of mirrors to a workpiece located a considerable distance from the source. And because the light is unidirectional, most of the output from the source can be collected by a focusing lens, focused to a very small spot, resulting in localized heating of a selected target material.

As described above, the process of stimulated emission leads to a light that is monochromatic, or consists of a single wavelength. By contrast, the light from a light bulb or from the sun consists of many wavelengths, including ultraviolet wavelengths that do not travel a great distance through air, all the visible wavelengths from violet to red, and some infrared wavelengths. The output from the common lasers used in industrial applications, the carbon dioxide and the Nd:YAG lasers, is monochromatic. That is, the energy output of the laser consists of light with a very narrow band of wavelengths. In both cases the central wavelength is in the infrared beyond the range to which the human eye responds. The wavelength of the Nd:YAG laser is at 1.06 µm and the wavelength of the carbon dioxide laser is further into the infrared, at 10.6 µm. This difference in wavelength has some important consequences, as will be discussed later.

4.1.3 Laser material interactions

A heated surface can lose heat that is carried away by three different means; conduction, convection and radiation. The element of an electric stove glows red hot, but cools very quickly when a pot of water is placed upon it, illustrating the loss of heat from the heating element by conduction. Since heated air rises, a hot water radiator heats the room because of the flow of air over its surface; this is an example of the convection of heat. The electrically heated filament of a lamp radiates visible light energy, which can illuminate a whole room.

When a focused laser beam is incident on a metal surface, a number of factors come into play. Obviously the incident laser energy will heat the surface on which it is absorbed. If one envisions that the tightly focused laser is a point source of heat on the surface, the temperature at that point is a balance between the rate at which heat is input at the surface, given by the power of the laser source and the fraction of the energy that is absorbed, and the rate at which heat is lost from the surface. At temperatures characteristic of the welding process, it is usually assumed that heat loss by radiation is negligible and that heat loss by convection through removal by the surrounding gases is a secondary effect. The primary means of heat loss is by conduction away into the metallic material being welded. Thus the temperature reached is a balance between the laser power input and the rate of heat conduction.

The temperature at the surface can reach the point at which the metal liquefies. The liquid pool resolidifies when the source of heat, the laser beam, is removed and heat is distributed by conduction through the solid material surrounding the liquid pool. If the laser was incident near a joint between two different pieces of material, both of which melt due to incident energy, a join or weld is established between the two pieces of material when the molten material solidifies.

Note that the molten pool is not stagnant, but is stirred rapidly. The primary force that causes the motion of the liquid pool is known as the marangoni force and is related to the surface tension. The fluid flow is controlled by the spatial variation of surface tension that exists on the weld pool surface. The surface tension gradient arises from the spatial variation in surface temperature and the temperature dependence of surface tension. The spatial variation of surface tension causes the molten metal to be drawn along the surface from the region of lower surface tension to that of higher surface tension and this may result in very large surface flows (Zacharia *et al.*, 1990). For pure metals and alloys the temperature coefficient of surface tension $d\gamma/dT$ is negative. Thus the surface tension is highest near the solid–liquid interface, where the temperature is lowest. The flow of the liquid pool is outward and away from the center of the pool. For metals with impurities, flow in the opposite direction may occur. The role of impurity elements and the spatial distribution of the laser energy in influencing the flow of liquid in the molten pool has been extensively investigated.

As the heat input is increased, the temperature increases until the vaporization temperature of the metal is reached. The laser beam drills a hole through the liquid pool, a hole which is filled with metal vapor. The laser passes through the metal vapor, contacts the liquid at the bottom of the hole and continues the drilling process. If the beam is moving with respect to the metal surface, the drilling process is not destructive, but forms a weld. As the beam moves, it continually melts more material at the front of the hole in the material. The molten material moves around the side of the beam and resolidifies at the rear of the hole. The flow of material is primarily in the liquid phase rather than in the vapor phase (Dowden *et al.*, 1983).

4.1.4 Welding modes

There are two different modes of welding. The first is designated the conduction mode, in which the size of the weld pool is limited by the conduction of the heat away from the point that the beam impinges on the workpiece surface. This mode of welding can be produced by either pulsed or continuous beams. If a pulsed beam is used, the molten pool and hence the weld nugget produced on a flat surface or a butt joint is approximately hemispherical. A repetitively pulsed beam can be used with a moving part to produce a series of overlapping weld nuggets that form a hermetic seal.

The second mode of welding is called deep penetration or keyhole welding. It occurs when the beam is intense enough to cause a hole filled with metal vapor to occur in the workpiece surface. It is generally considered that a laser power of one MW/cm^2 is required for keyhole welding in steel workpieces. A somewhat higher power is required for aluminium workpieces. Depending on the welding conditions, the hole may extend either part way or entirely through the workpiece. This mode of welding is most often performed with a continuously operating laser, although there has been some work done with repetitively pulsed or modulated beams. The welds produced by the deep penetration mode of welding have a high aspect ratio; that is, they are relatively deep and very narrow. Aspect ratios as high as 10 to 1 are not uncommon.

4.2 Energy efficiency

4.2.1 Energy conversion

A laser is an energy conversion device. The industrial laser converts electrical energy to light energy, but it does not do this very efficiently. A carbon dioxide laser has a wall plug efficiency (that is, the energy output, measured in watts of average power, divided by the electrical energy drawn from the wall) of less than 10 %, and Nd:YAG lasers have wall plug efficiencies of 1 or 2 %. The remainder of the electrical energy input is carried away by the flow of cooling water. In spite of these inefficiencies, the beam can be delivered very efficiently to the workpiece. Newer generations of lasers such as diode lasers have a considerably larger energy efficiency than the carbon dioxide and Nd:YAG lasers currently widely used.

Because of the low energy efficiency, it is important that the energy be used expediently when applied to the weldment. There are two energy efficiencies that are important. The first is the energy absorption efficiency, also called the energy transfer efficiency, or arc efficiency in the case of an arc welding process. This designates the fraction of the incident beam energy that is absorbed in the workpiece. The second energy efficiency is designated the melting efficiency and is characteristic of what happens to the energy once it is absorbed in the workpiece.

4.2.2 Energy absorption efficiency

The energy absorption efficiency is the fraction of the laser energy directed at the workpiece that is absorbed into the workpiece. Two reasons why incident energy may not be totally absorbed by the material are reflection from the workpiece and transmission through the workpiece. Some of the energy may be reflected from the surface and not absorbed. The absorption of laser radiation into metals depends on the nature of the metal, the temperature, the wavelength of the laser, the roughness or surface condition of the metal, and the angle of incidence of the radiation onto the material. The wavelength of the Nd:YAG and diode lasers result in greater absorption into most metallic materials than that of the carbon dioxide laser. As temperature increases, however, so does the absorption of the laser radiation by the material. Thus, even in a material that is largely initially reflective, as long as the material absorbs part of the energy of the beam and starts to heat, a larger fraction of the beam is absorbed and the heating process accelerates. If a keyhole is formed in the material, the keyhole acts as a trap and most of the beam is absorbed.

If the keyhole extends completely through the material, some energy may pass through the bottom of the keyhole. In the optimized welding process, this energy loss is minimal. A more serious energy loss is from absorption of laser light in a plasma which may occur above the material surface, or light scattering from a plume above the surface. The plasma consists of ionized gas which absorbs energy from the laser beam by a process known as inverse bremstrahlung absorption (Offenberger *et al.*, 1972). The absorption increases with the square of the wavelength and hence is considerably more severe for the carbon dioxide laser than for the Nd:YAG laser. When welding with carbon dioxide lasers with power greater than a few kilowatts, a plasma suppression jet directs a flow of helium gas at a slight angle to the surface, immediately above the keyhole region. The jet is usually positioned to blow the plasma, and other gas coming from the keyhole, onto the cold surface immediately ahead of the weld area. This has been shown to limit the loss of easily vaporized alloying elements from the weld metal (Blake and Mazumder, 1982) and also may serve to preheat the metal ahead of the beam.

The net absorption of laser energy (or energy from an arc welding source) into the weldment is most accurately measured by a Seebeck calorimeter. This is an insulated box with an insulated lid that is manually shut immediately after the weld is completed. The box contains thermocouples that measure the total flux of heat through the walls of the box during its return to room temperature; by integrating the flux of heat, the amount of heat absorbed by the metal can be evaluated. Measurements of absorption efficiency have been presented by Banas (1986) for a variety of different materials.

4.2.3 Melting efficiency

The melting efficiency is the fraction of the energy absorbed by the material that is used actually to melt the metal to form the weld. Measuring this quantity requires the use of instrumentation such as the Seebeck calorimeter to determine the amount of energy absorbed by the material. The amount of energy to perform the weld is determined by examining cross-sections of the weld to determine the area, knowing the speed of the weld, and using data for the heat capacity for the material between room temperature and the melting point and the heat of fusion. Usually several cross-sections are taken, and an average calculated, to account for possible fluctuations in the material or the process.

The melting efficiency is a consequence of the heat flow patterns in the material. Many conduction welding processes can be thought of as originating from a point source moving across the surface of the material being welded. In this case, heat can be conducted in three dimensions away from the point source, in the direction of motion of the heat source, perpendicular to the direction of motion, and into the depth of the material. On the other hand, a deep penetration weld can be thought of as originating from a line source of heat, extending through the thickness of the material. In the latter case, heat conduction away from the heat source is only two-dimensional as there is already a distribution of heat through the thickness of the material. Thermal losses are less severe in the latter case since there are fewer dimensions in which the heat can be conducted away. Some thermal losses are inevitable; it is impossible for one area to be heated to above the melting point without some heating of the surrounding area. It has been shown that the maximum possible melting efficiency for welding is 37% for three-dimensional heat conduction and 48% for two-dimensional heat conduction (Swift-Hook and Gick, 1973).

4.3 Laser parameters: their measurement and control

4.3.1 Laser power and power density

Continuously operating or CW lasers are invariably rated by their power output, measured in watts or kilowatts. This rating refers to the power generated at the output mirror or window of the laser and is usually measured by an internal power meter and displayed on a monitor. For some cases, some fraction of the beam is split from the beam delivered to the workpiece and monitored on a power meter, so there is always a display present. In other cases, a power reading can be taken only when the beam is not delivered to the workpiece; that is, the beam either goes to the power meter or to the workpiece, but not to both simultaneously. Note that high power laser power meters are usually thermal; the beam causes something to heat and the rate of heating of the object is related to the energy-incident beam. Because thermal conduction is a relatively slow process, such power meters have a slow response time, usually seconds or longer. These meters would not detect modulation on the beam or start-up transients, which may have a deleterious effect in welding.

It is not the power at the laser, but the power at the position of the workpiece that is of interest in the welding process. Although gold-coated mirrors may have a theoretical reflectivity of 99%, it is usually assumed that there is a 4% loss on each mirror surface. In a multi-axis motion system, there may be a number of mirrors and lenses in the beam path between the laser and the workpiece. Consequently, the power at the position of the workpiece can be considerably lower than the laser output power and should be verified and used in any data records of the process. Similarly, with lasers in which the beam is delivered by fiber optics from the laser to the workpiece, there may be losses in introducing the beam into the fiber and extracting it for focusing on the workpiece.

For welding with pulsed lasers, the relevant parameter is not the power but the energy per pulse, which is similarly diminished between the laser output and the workpiece. The weld is performed by individual pulses. The average power with pulsed lasers is proportional to the pulse repetition rate, for a given laser pulse energy and determines the rate at which spot welds are performed on the workpiece. The dimensions of the weld are determined largely by the energy of individual pulses, however.

Whether using CW or pulsed lasers, however, the power and pulse energy are not as important as the power density, measured in units such as watts per square centimeter, or the energy density, measured in joules per square centimeter. To determine these quantities one needs a method of measuring the size of the beam, as discussed in a later section.

4.3.2 Laser modes

The term mode describes the distribution of laser intensity within the beam. For industrial lasers, the term is short for 'transverse modes' since the other type of modes, longitudinal modes, are relevant only to lasers used for precision sensing. The transverse mode, or distribution of intensity in the plane perpendicular to the optic axis, is determined by the nature of the mirrors used in the laser construction. There are four types of laser modes: stable, unstable, waveguide and hybrid stable–unstable.

Any light beam, by its very nature, tends to spread out or 'diffract' as it passes through space. A stable mode is formed when the light radiation bouncing back and forth between two mirrors of the laser is refocused when one or both of the mirrors has a curved surface. The refocusing counteracts the tendency of diffraction to spread the beam out and confines the beam near the axis of the two mirrors. One of the mirrors must be partially transmitting, to allow some fraction of the beam to emerge from the laser to perform useful work. The remainder of the beam is reflected back into the laser, is amplified by the medium of the laser to compensate for the amount lost by transmission through the mirror, and is retroreflected by the second mirror. The two retroreflecting mirrors are said to form a 'resonator', because the amplified light resonates between them. The stable resonator is one in which the curvature of the mirrors is such that the light is confined to near the axis defined by the two retroreflecting mirrors.

There are a number of transverse modes, or distributions of laser radiation, that can be formed by a stable resonator. The modes are solutions of the mathematical equations which describe the propagation of light, with the boundary conditions established by the two resonator mirrors. The preferred mode is one that is strongest along the axis, with the intensity decaying in a Gaussian fashion with distance away from this axis. This is called the TEM_{00} mode. The other modes, or solutions of the mathematical equations, can also be realized in practice. High power lasers often operate in a multimode fashion, with a variety of the modes operating simultaneously.

The intensity distribution in the TEM_{00} beam is circularly symmetric and is given by

$$I(r) = P/(\pi\omega^2) \exp(-2r^2/\omega^2) \qquad [4.1]$$

where r is the transverse distance from the optic axis, P is the total power in the beam and ω is the beam radius. Owing to diffraction, the beam expands as it propagates through space; however, one property of the Gaussian beam is that it remains Gaussian as it propagates. Therefore the propagation of the TEM_{00} beam can be described by the way that the beam radius changes, which is

$$\omega(z) = \omega_0^* \operatorname{sqrt}(1 + (z-z_1)^2/z_0^2) \qquad [4.2]$$

Here ω_0 is the minimum value of the beam radius, designated the 'beam waist', which occurs at a position z_1. Both ω_0 and z_1 are determined by the nature of the retroreflecting mirrors and the distance between them. If one of the mirrors is flat, rather than curved, the beam waist will often occur at the position of the flat mirror. $z_0 = \pi\omega_0^2/\lambda$ describes the expansion of the beam as it propagates, where λ is the wavelength of the laser light. For $z-z_1 \gg z_0$, then the beam radius expands linearly with distance $\omega \sim \theta z$ where

$$\theta = \omega_0/z_0 = \lambda/\pi\omega_0 \qquad [4.3]$$

is the half-angle divergence (that is, the divergence of the radius, rather than of the diameter, of the beam).

It is difficult or impossible to find optical materials that are partially transmitting that can survive the high power beams. Consequently, the optical system for high power lasers allows the laser beam to expand as it reflects between the two mirrors; the distribution of intensity becomes 'unstable' with respect to confinement along the optic axis. A laser output arises not because one mirror is partly transparent, but because either the beam gets so large it spills over the edge of the smaller of the two mirrors, or a 'scraper' mirror extracts the outside portion of the beam, allowing the inside portion to be reflected back into the amplifying medium. Unstable resonator beams are described by a factor called the magnification factor M, which is the outer diameter of the beam divided by its inner diameter. The magnification factor is determined by the amount of curvature of the two mirrors that form the laser cavity and the distance between them. Most carbon dioxide lasers with unstable resonators produce their maximum output when the magnification is near 2. Use of optics that give a larger magnification result in a sacrifice in output power.

The waveguide resonator is utilized in diffusion-cooled lasers, where there is no convective cooling of the laser gas mix. Instead, the electrically or RF excited gas is cooled by conduction through the gas mix to the water-cooled walls of the excitation volume. Effective cooling requires the laser geometry to be kept small and the shape of the laser output is determined by the 'guiding' of the radiation between the walls of the chamber as much as it is by the curvature of the reflecting mirrors.

A hybrid stable–unstable laser is one which has a stable resonator configuration in one direction (e.g. the *x*-direction) and an unstable resonator configuration in the *y*-direction, where the axis of the laser beam is along the *z*-direction. One method of scaling of gas laser excitation to higher power is to extend the electrodes in the direction of the transverse gas flow; this leads to an excitation region that is rectangular shaped. A hybrid stable–unstable resonator has been investigated as one way of extracting energy from the rectangular excited region. More recently, the diffusion cooled laser has been extended to a 'slab' geometry, in which two extended electrodes are placed on either side of the rectangular shaped excited region. In this case, the laser resonator results in waveguide laser modes in the direction between the electrodes and a stable mode in the direction perpendicular to this. Special optical arrangements are used with these lasers to produce a beam of high quality and minimum divergence. They are used for welding applications involving sheet metal and polymers.

4.3.3 Beam characterization

The beam width is defined as the diameter of a circle that includes $1 - 1/e^2$ ~ 85% of the total power of the beam. For the TEM_{00} beam described above, this definition of beam diameter corresponds to twice the beam radius ω.

The quality of a beam is a measure of its ability to be focused to a small spot size, raising the intensity, or power per unit area (or energy per unit area, for pulsed beams) to a high value to do useful work. The quality of the beam is determined by the resonator design and the choice of the retro-reflecting mirrors. The measurement of beam quality is called the M^2 factor. Note that this is a different M from the one mentioned above for the magnification of an unstable resonator. European laser practitioners use a K factor that is related to the M^2 factor by $K = 1/M^2$. For the lowest order Gaussian mode, $M^2 = 1$. Most laser suppliers list the M^2 value for their laser. The M^2 can be determined by determining the beam waist ω_B of the laser, via multiple measurements of the beam radius, and the divergence of the laser. The quality factor of the beam is then given by the ratio of the beam divergence to the value that the beam divergence would have if it was the TEM_{00} mode; that is,

$$M^2 = \theta_M/(\lambda/\pi\omega_B) \qquad [4.4]$$

where θ_M is the measured value of the half-angle beam divergence.

Why is the beam quality important? A low quality beam diverges more rapidly than a high quality beam, and is focused to a minimum beam radius a factor of M^2 larger than a similar low order mode can be focused. This means the intensity at the focus is lower by a factor of M^4 than that of a similar TEM_{00} mode and the ability of the laser to do work is diminished.

The depth of focus, or the range of distance over which the beam maintains a minimum value, is also smaller by a factor of M^2. Consequently, the criticality of maintaining focus in a welding operation is more severe with a beam with a higher order mode.

Note that the beam quality is determined by the characteristics of the laser. As the beam propagates through space, if it is enlarged or focused by perfect lenses, its quality remains the same. If the beam passes through imperfect focusing systems, its quality can be worsened. In some cases, beam quality has been improved by focusing the beam through a pinhole and recollimating the transmitted beam. The pinhole absorbs the fringes of the beam that represent power in higher order modes.

4.3.4 Measurement of focal spot size

The beam diameter at the focus, also called the spot size, determines not only the fineness of the features that can be cut or welded, but also determines the intensity, or power per unit area, at the focus. Laser material interactions are determined by the intensity, hence the focused beam size is a very important parameter.

To the zeroth order approximation, the beam radius is controlled by the quality of the beam, as reflected in the divergence θ of the beam prior to the beam striking the lens:

$$\omega_f \sim \theta F \quad [4.5]$$

Here F is the focal length of the lens or mirror used to concentrate the beam on the workpiece. However, an actual measurement of the power distribution in the region of the focus provides a more direct and reliable source of information. This actual measurement takes into account any aberrations that may be produced by the focusing lens. Since most metalworking lasers have the power to vaporize any known substance, measuring the focused spot size of the laser is a challenging operation. It has been achieved using a variety of commercially available equipment based on light scattering or light collection by a rapidly scanning wire or hollow needle through the focal volume. The spinning needle survives since it does not spend enough time in the focal region to accumulate sufficient heat.

From the signal received, the beam radius can be calculated. If the detection system is moved along the direction of propagation through the focus, the beam size can be found as a function of position. Then the minimum beam waist can be found and the beam divergence at positions away from the beam waist. From this data, the beam quality can be evaluated using Equation [4.2], where $z_0 = \pi \omega_0^2 / \lambda M^2$ is the parameter that describes the expansion of the beam away from its minimum value.

Using this method, it was found that the output of a Mitsubishi 1.6 kW carbon dioxide laser, focused by a 6.3 cm focal length lens, had a minimum radius of 0.33, 0.35 and 0.62 mm if the laser was operated at 200, 800 or 1600 W, respectively. The measured beam quality factor was $M^2 = 2.6$, 2.4 and 4.7 at the three power levels, indicating that the beam quality at the workpiece degraded as the power level was increased. The intensity, or power per unit area, was actually higher if the laser was operated at 800 W than if the laser was operated at 1600 W. It is likely that at least some of the beam degradation was due to distortion of the focusing lens due to heating as the zinc selenide lens material absorbs a small amount of energy from the beam. This behavior of the beam would have been very difficult to ascertain without actual measurements of the beam spot size.

4.3.5 Parameters of pulsed lasers

The average power output of pulsed lasers represents the average power delivered by the laser; for example, if the laser delivered pulses with 7 J of energy at a repetition rate of 10 Hz, the average power is 70 W. If the pulses last a millisecond, often the peak power is calculated by dividing the pulse energy by the duration of the pulses. For example, if the 7 J pulses last a millisecond, the peak power is said to be 7 kW. But it is more correct to consider this the average power during the pulse; the pulse itself may have fluctuations that can only be observed with a fast detector and oscilloscope. Consequently, the peak power may be significantly higher than the average power during the pulse.

Most pulsed lasers used for metalworking applications at the time of writing are Nd:YAG lasers. The neodymium atoms in the yttrium aluminum garnet rod absorb energy from the flashlamps; some of this energy is extracted in the laser pulse, but the remainder is conducted through the YAG rod to the water-cooled walls. Consequently, there is a temperature distribution across the rod, which then behaves as a lens. The amount of lensing in the rod is believed to be determined by the average power input to the rod. Since the rod becomes a lens, it affects the propagation of light between the resonator mirrors with the result that the mode of the output depends on the average power to the rod. The focused spot size can be measured using the spinning wand or hollow needle technique discussed above, but the firing of the pulsed laser must be synchronized with the revolution of the rod (Graham and Weckman, 1995). The peak intensity at the workpiece surface does increase with average power output, but the increase is less than proportional to the increase in power output.

Equally important as the change in minimum beam waist of the focused beam is the shift in the position of best focus as power is increased. Clearly, the position of best focus should be found for every average power level for which the laser is operated, otherwise welding may be attempted with the beam considerably out of focus.

A later generation of Nd:YAG lasers used a rectangular solid piece of crystal, with the beam redirected through the excited medium in such a way that any thermal distortion of the beam would be cancelled out. This style of laser was more expensive, however, and found limited market acceptance. The output of many Nd:YAG lasers is delivered by a fiber optic to the focusing head, which concentrates the beam on the workpiece. It is widely reported that the fiber 'homogenizes' the beam, and that the spot size on the workpiece is the image of the end of the fiber. The implication of these reports is that any energy-input dependence of the mode produced by the Nd:YAG laser is not important when the beam is fiber delivered. This is not necessarily true. Boechat *et al.* (1993) have reported that the length of fiber required for the output beam to be independent of the launch conditions is far longer than the normal fiber lengths used in welding lasers. Moreover, recent measurements showed a dependence of the measured spot sizes of fiber-delivered lasers on the laser operating conditions (MacCallum *et al.*, 2000).

4.3.6 Other beam-related factors

As mentioned above, there was a shift in the position of the focus as a function of laser power when using an Nd:YAG laser. Similar effects have been noticed with carbon dioxide lasers. For example, with a 20 kW CO_2 laser focused by a 70 cm focal length mirror, a shift in position in excess of a centimeter was observed between spot size measurements undertaken at 2 kW and at 20 kW. The shift, which was accompanied by a change in focal spot size, is attributed to beam-induced thermal distortion of the window of the laser chamber.

Thermal distortion of the focusing lenses or output windows does not occur instantly. It was ascertained that the focus position of the output of a 1.6 kW carbon dioxide laser shifted in a time period of the order of 60 seconds after turn-on. Similar shifts were observed with an Nd:YAG laser, with 120 s required for stabilization. This shift was attributed not to warm up of the rod, which ought to take place on a time scale of a few seconds, but to temperature changes in the entire water cooling circuit.

It is assumed that the laser reaches its programmed power instantaneously after the command is given. This was investigated on the 1.6 kW laser described above. Measurement of the turn-on transients using a non-thermal detector with a fast response time showed that at times the laser overshot its programmed value by 25%, settling down to the steady state value in approximately a 10 s time period. Repeating the measurements several days later, however, the power was observed to increase gradually to the programmed value over a 10 s time period. The reason for the different response in the two cases was not ascertained. This laser was sealed, but the gas was replenished on a weekly basis. Possibly the different response was related to the age of the laser gas. Regardless of the reason for the time variation, successful welding with this laser was achieved only when the laser action was activated with the beam directed off the workpiece, into a beam dump, and welding commenced after a sufficient time period for the power to stabilize. This time dependence was not displayed on the power meter built into the laser and the laser had been successfully used for cutting operations for many years. Since significant energy input for the cutting operation comes from the heat of reaction of steel with the oxygen assist gas, the cutting operation is possibly less dependent on the laser power level than is the welding operation.

4.3.7 Other parameters

In controlling or specifying a laser-welding process, a number of other parameters affect the process. Welding procedure specifications generally class these as 'essential variables' and 'non-essential variables'. An essential variable is one that can have a major impact on the weld quality; if a substantial change is made in an essential variable, the welding process must be requalified, which may be an expensive and time-consuming procedure. Essential parameters include, for example, the laser power (or pulse energy, pulse duration and pulse repetition rate for pulsed lasers), beam mode profile, lens focal length, focal point position, raw beam size, motion speed, number of passes, angle of incidence, welding position, nozzle gas type or composition, auxiliary gas type or change in composition, backing gas type or change in composition, plume reducing gas jets including orientation, flow rate or pressure of various gases, change in material or in filler metal type or size, joint design and joint gap.

Gas shielding is usually used for laser welding. With Nd:YAG welding, argon is the preferred shield gas, as it is heavier than air and falls onto the workpiece. In welding with the carbon dioxide laser, however, helium is normally used as it has a higher ionization potential than argon. Plasmas that absorb and scatter the laser radiation are more easily created at the carbon dioxide laser wavelength than at the Nd:YAG laser wavelength. Several different types of gas shields may be used. In a weld that completely penetrates the workpiece, a backing or underbead shield may be necessary. For very high speed welding, an auxiliary or trailing gas shield may be necessary. For spot welding, a simple gas flow through the nozzle may be sufficient; this serves not only to protect the weld from contamination, but also to protect the lens from fumes. For high power carbon dioxide laser welding, a plasma suppression jet may be required to blow the plasma out of the beam path. Gas flow must be carefully controlled. For cost purposes it is desirable to keep gas flows low, and indeed too high a gas flow may aspirate air into the gas stream resulting in weld contamination. Gas flows should be high enough, however, to provide adequate shielding in spite of random drafts.

Note that some steel types can be welded in air, without an inert gas shield. Other materials, such as titanium, which are often arc welded inside sealed and purged chambers can be successfully laser welded in air with only a nozzle gas shield. This is because the spot welds produced by a repetitively pulsed Nd:YAG laser are often fairly small and cool rapidly so there is no hot weld pool extending beyond the region covered by the gas flow.

4.3.8 Filler metal

Laser welding is most often carried out without filler metal; this process is called autogenous welding. There are two reasons why one might use a filler metal. As described above, the small size of the focused laser beam means that very good edge preparation must be used. However, the requirements for edge preparation can be relaxed if a filler metal is used. The second reason to use a filler metal is to control the metallurgy of the weld metal.

Filler metal can be applied in the form of wire, powder, or preplaced inserts such as rings or discs. The feeder for filler metal should be integrated with the laser control circuits, but note that many wire feeders are not sophisticated enough to produce reliable welds. In particular, at the end of weld, the wire feeder should be turned off a short interval before the laser beam is. Otherwise, the wire will freeze into the weld pool requiring a manual operation to free it. Moreover, at the start of the weld the position of the wire must be carefully set and the advance of the wire integrated with the laser turn-on time. Co-ordination between the feeder and the laser control circuits is less critical when using a powder feeder rather than wire feeder, as discussed in a later section on laser cladding and weld repair.

4.3.9 Positioning of the beam

Laser welding is usually carried out with the laser beam directed at the seam between the two parts to be welded. There are two reasons for welding off the seam. The first is to control the metallurgy of the weld metal when, for example, welding a low carbon steel to a high carbon steel. This would happen when welding a formed component to a machine component, for example. To prevent cracking, it is beneficial to attempt to lower the carbon content of the weld metal. The position of the bead is judiciously located so the major part of the weld metal originates with the low carbon steel, in such a way that the total weld penetration is unaffected.

A second reason for laser welding with the beam positioned away from the seam between the two materials being welded is to enhance the absorption of the beam. For example, 12 mm thick copper which is normally highly reflective to the output of a carbon dioxide laser, has been successfully welded to 12 mm thick nickel with a 9 kW laser beam by locating the beam approximately 0.25 mm onto the nickel side of the seam.

4.4 Weld quality assurance

4.4.1 Introduction

There are four types of observations of weld quality: visual examination, destructive analysis, non-destructive examination and *in-situ* observations. These are discussed separately below. Which types of weld inspection procedures are used may depend on the weld procedures, the weld qualification procedures, and/or codes for welding. For example, all welds on pressure vessels, or on valves, gauges and fittings that are used on pressure vessels, are subject to Section IX of the American Society of Mechanical Engineers (ASME) welding code, or local Boiler-and-Pressure Vessel codes.

4.4.2 Visual examination

A visual examination gives information about the surface regularity of the weld and the presence or absence of cracking, surface porosity or undercuts. Restrained welds in higher carbon steels have a tendency towards cracking due to the hardness of the weld metal and heat-affected zone, particularly at the weld close-out. The width and bead profile of welds can be compared with that observed in trial welds under similar conditions taken before the commencement of production operations. If the weld is fully penetrating, and there is access to the underbead, the presence or absence and the regularity of the underbead ensures that complete penetration did occur. Visual examination by the welder or weld machine operator is often the first step in ensuring the correct operation of the equipment and ensuring there is no variation in the preparation of materials, cleanliness or shielding gas.

4.4.3 Destructive analysis

Destructive analysis is the process of cutting up the weldment to observe the depth and shape of the weld metal and the heat-affected zone. Often the material pieces containing the weld are mounted in a standard size mount of epoxy or plastic, metallographically polished and chemically etched to show up microstructural features. The section of the weld that is displayed can be a cross-section of the weld or can be longitudinal along the length of the weld. This process provides the best possible indication of the quality of the weld in the location of the cross-section; however, it does destroy the welded structure.

In some production operations, the destructive analysis is carried out on a sampling of parts (e.g. 1 in 50, or 1 in 1000), is carried out on parts that are rejected for some other reason unrelated to the welding operation, or is carried out on less expensive simulated parts that would have a similar heat flow pattern to the real weldment. The destructive analysis is time consuming and expensive, in that work pieces are destroyed, and is therefore used judiciously. Nevertheless, there is no substitute for the destructive analysis during the establishment and qualification of welding procedures, and the calibration of non-destructive analysis and visual analysis equipment.

More sophisticated destructive analysis can include hardness testing, tensile testing, bend testing, dynamic-tear, impact and fatigue testing of the weld metal and heat-affected zone. Tailor blank welding for the automotive industry utilizes a cupping test, where the material including the weld must be significantly deformed with a semi-circular punch, without breaking the weld metal. Structural welds for submarines are subjected to an explosion bulge test. Here plates containing a weld are deformed by an explosion to produce a significant bulge which the weld must survive.

4.4.4 Non-destructive examination

Non-destructive examination (NDE) procedures include the use of X-rays, ultrasonic examination and acoustic emission monitoring. X-rays can detect weld porosity, lack of side-wall fusion, missed seam defects and inclusions. In some cases, for example in the nuclear industry, 100% inspection of parts is required by welding codes. Ultrasonic inspection can detect the boundaries of the weld material and porosity that is larger than the ultrasonic wavelength, and a missed seam if the seam is perpendicular to the direction of propagation of the ultrasound. Often a hermiticity check, using, for example, a helium leak detector, is used in production environments.

4.4.5 *In-situ* observations

A skilled manual welder, using a hand-held welding torch, carefully observes the size and position of the weld pool, the light emitted from the weld pool and the sound emitted from the arc. On the basis of these observations, he or she judiciously adjusts the speed and position of the torch, and perhaps the power of the arc. It is the intent of *in-situ* observation equipment to duplicate and extend the operations carried out automatically by the skilled worker. In the last few years, there have been extensive investigations of various visual and audio signals from a laser weld pool with the aim of using these signals for detection of weld quality and ultimately for the control of the welding process.

Electronic signals have been derived from the incident laser light that is reflected from the weld pool, the ultraviolet emission from the weld pool, the infrared emission, the audio emission carried through the atmosphere and ultrasonic emission carried away from the weld pool by the weldment itself. In some cases, the light emitted from the weld pool has been spectroscopically analyzed to indicate shield gas contamination and the sound emitted from the weld pool has been frequency analyzed. There were indications that the dominant frequency in the sound emitted may be related to the depth of the keyhole, in much the same way that the dimensions of an organ pipe control the notes that are emitted. These observations have not yet resulted in wide acceptance of such technology as a control measure. The types of observation best suited for control may depend on the materials being welded and the geometry of the weldment.

4.5 Advantages of laser beam welding

Metzbower, in 1981, presented a review on laser technology for thick-section welding, as the technology had been developed up to that time. His paper showed data on welding of the same type of steel with four different welding processes. A butt joint between two pieces of 2.7 mm thick HY-130 steel was welded with an 11 kW laser beam at speeds between 12.7 and 16.9 mm/s. A similar weld was produced with a 40 kV electron beam at 21.2 mm/s, but this weld showed undercutting and required a defocused cosmetic pass on both sides of the plate of steel. The heat input to the part for the laser weld was 0.7 kJ/mm, whereas for the electron beam weld it was somewhat less, about 0.44 kJ/mm.

The same weld was done by shielded metal arc (stick) welding (SMAW). The joint preparation of the material was a 60 degree groove and a 120 °C preheat was applied. The welding speed was 3 mm/s with a 125 A arc at 25 to 30 V. This corresponds to a heat input of 1.1 to 1.18 kJ/mm per pass, considerably higher than that of the two deep penetration welding processes. In the SMAW process, seven passes were required to fill the joint completely. The net welding speed used in the laser beam process was 30 times faster than that of the shielded metal arc weld.

Gas metal arc welding was also used to make the weld on 12.7 mm thick HY 130 steel, also with a 60 degree groove weld preparation and 120 °C preheat. The welding speed was about 6 mm/s with a 300 A, 24 V arc. Five passes were required, with each pass having a heat input of 1.1 kJ/mm. The welding speed in the laser process was considerably faster.

The above results typify the advantages of laser beam welding. Thick material can be welded at high speed in a single pass. The narrow width of the deep penetration weld means that the heat input to the weldment is considerably smaller than in the arc welding processes. The laser beam weld had similar characteristics to the electron beam weld, but the electron beam welders require a vacuum system since the beam cannot be transmitted through the air. The lower heat input of the high energy beam welding processes significantly reduces the welding induced distortion and may have beneficial metallurgical consequences.

Repetitively pulsed welders have had considerable success in the medical electronics industry and the electronics industry. Nd-YAG lasers that produce a well-defined energy per pulse at a programmed pulse repetition rate are ideal for welding of thin metal parts; the laser energy can be chosen to be just sufficient to join the two pieces of metal together without overheating or damaging internal components. Many of these lasers monitor the laser pulse energy and have internal feedback circuits to keep the laser energy at a predetermined level.

The output of the Nd:YAG laser is often delivered by fiber optics to the workstation. Since the fibers are essentially loss-less, this means that the laser does not have to be physically close to the welding operation. In many cases, electronics components are assembled in clean rooms or dry rooms. These rooms are expensive to build and maintain; consequently, it is advantageous to have the laser physically outside the room, with the output delivered inside the room by fiber optics.

4.6 Suitability of laser beam welding

Laser welding is an accurate and fast process. The reason that it is relatively fast is that the fusion zone is relatively small. A deep penetration weld heats only the seam and a small area around it, rather than the large area that is welded when using a deep V-groove preparation and arc welding. Sheet metal can be welded rapidly by focusing the beam to a spot size of the order of the thickness of the metal. This accuracy has a disadvantage, however. Weld joints have to be precisely prepared as the process will only tolerate a very small gap between the two parts to be welded. For thick materials, a maximum gap of 3% of material thickness is quoted (O'Brian, 1991). If a wide gap is used, a weld with an underfill or an undercut results, or, in some cases, part of the beam is transmitted through the gap and is not available for the welding process. Consequently, the laser beam process is used when the expense of a machining process or a precision weld joint preparation process can be tolerated.

For example, coil joining in the steel industry and tailor blank welding in the automotive industry were only successful after a careful examination of the shearing process and optimization of the shearing to produce edges suitable for laser beam welding. Here, the companies decided the benefits of laser welding were so great that the effort involved in developing and implementing methods to prepare high quality edges was justified. However, the need for precision edge preparation has kept laser applications out of many fabricator shop environments.

4.7 Process selection

In many cases, the selection of laser welding is justified on both financial and technical grounds. Laser welding equipment is considerably more expensive than is arc welding equipment. Moreover, the laser process is almost always carried out with automated equipment, which adds to the cost. For safety reasons, laser welding is usually carried out inside a safety enclosure, which is an impediment to loading and unloading weldments rapidly. Businesses will generally only invest in laser welding equipment if they foresee multi-year production runs or sequences of production runs lasting at least as long as the equipment is being depreciated.

The welding process will always be carried out using a process that involves less expensive capital and operating costs unless there is some special reason to use the laser process. It is the low distortion nature of the welding process that has led lasers to be the preferred method for laser welding of gears for automotive transmissions. It is for metallurgical reasons that lasers are the preferred method for weld repair of the tips of blades from gas turbine engines. The speed of the process is the reason it has been adopted for pipeline welding on offshore pipe-laying platforms. The platforms are reputed to cost a million dollars a day to operate, so faster welding saves money.

4.8 Current laser beam welding applications

The most recent attempt to survey comprehensively laser applications was performed by the Electric Power Research Institute a number of years ago and is severely out of date (Brushwood, 1984). More recent surveys have relied on discussions with sales representatives from different laser manufacturers and are often incomplete because of considerations of company confidentiality.

The widest acceptance of laser welding applications is in the automotive industries. Laser welding is the method of choice for welding of components of gears; perhaps the largest concentration of high power lasers in the world is near Kokomo, Indiana in the United States, where there are three large automotive transmission plants each with multiple laser welding systems. It has been said that the advent of low heat input welding processes resulted in dramatic changes in the design of gears. Whereas previously a large gear may have had to be machined from a large block of steel, the laser welding process allowed the gear to be made, for example, from a stamped or formed flange welded to a machined hub.

Another area in which lasers have found applications in the automotive industry is that of tailor welded blanks. After welding, the blanks are formed into auto-body parts such as door panels, and holes are cut in the formed parts as needed. This advanced technique enables designers to tailor or optimize the materials in their parts, while keeping the overall weight of a part to a minimum. If a particular location in the part requires a particular type of steel for reasons of strength or corrosion resistance, then a piece of that type of steel is welded to the part while it is still a flat sheet, before being formed into the various complex shapes. The result of instituting a design-for-manufacturing program is reduction in the number of parts and assembly time required per vehicle. Experience has shown that over 20 parts can be eliminated by the use of tailor welded blanks for door parts and frames in a single vehicle. The laser welding process produces parts of superior quality and offers significant advantages over the traditional spot welding operation. The continuous butt weld of the tailor welded blank replaces the discontinuous joint that would result if the parts were joined after forming. Greater dimensional control is achieved. The continuous laser weld eliminates the need for sealant, in addition to achieving greater strength while reducing weight. Automobile manufacturers have adopted tailor welded blanks not only as a cost saving, but also to reduce weight and hence increase fuel efficiency to satisfy legislative requirements.

Other manufacturers of products such as electronic cabinetry and household appliances are also investigating the production of goods from tailor welded blanks. The ability to manufacture tailored blanks can best be utilized by an evolution in design philosophy. In a mass production scenario, engineers should learn to design parts based on raw materials that are optimized for a particular application. Successful implementation of tailor welded blanks requires the development of high-speed, high-quality laser welding processes, producing a minimum of overbead that influences the subsequent punch-and-die forming operations. The development of the laser welded tailored blanks was accompanied by developments in high accuracy shearing processes, producing a smooth edge without any further processing prior to welding.

After a slow 15 year incubation period, laser welding of automobile bodies has been successfully implemented. One of the first applications used the hybrid or laser-assisted arc process with a gas metal arc weld used in conjunction with the laser weld. The filler metal added using the arc process relaxed the tolerances required to make a good weld. Within two or three years of this first installation, one manufacturer was said to be using 240 lasers for body welding.

Another high profile laser installation occurred in the 1980s when Kawasaki Steel Co. implemented laser welding for coil joining (Kawai *et al.*, 1984). By joining the coils of steel produced in the steel mill together to produce one long strip, subsequent processing through the cleaning and chemical cleaning process was simplified. The manufacturing process was essentially turned from a batch process into a continuous process, eliminating the need to feed new strips of steel continually through guide coils. Five kilowatt lasers were used, but auxiliary equipment included a high accuracy shearing mechanism and an abrasive wheel grinder. The grinder allowed ease of transfer of the material over subsequent rolls. An auxiliary wire feed lowered the carbon content of the weld metal when the system was used for joining high carbon steel.

Another instance of using laser joining to turn a batch process into a continuous process was in the application of 45 kW continuous lasers at the Ohita plant of the Nippon Steel Corporation (Anon, 1995). The laser welded rough rolled hot slabs to each other, allowing continuous roll finishing, thereby achieving 20% higher productivity. It was anticipated that the new process would produce thinner gauge hot steel plate and formability would be improved. In addition, high power lasers have been successfully implemented into pipe mills, welding steel pipe with wall thickness up to 16 mm (Ono *et al.*, 1996).

Repetitively pulsed Nd:YAG lasers, with a close spacing between repetitive pulses, provide a hermetic seal. This ability has been used in sealing of batteries, pacemakers and relays since the early days of the laser (Bolin, 1976, 1983; Fuerschbach and Hinkley, 1997). Applications in the electronics industry include components of electron guns and grids for televisions (Notenboom, 1984), thermocouples, ink cartridges for fountain pens, relays, telephone switching gear, microwave components, lamp electrodes, gyroscope bearings and valve components (Bolin, 1983). A particularly challenging mass production job is the sealing of glass-to-metal feedthroughs into electronic components; the possibility of cracking the glass due to excessive heat input is avoided because of the fine control in the laser welding process.

4.9 Related processes

Laser cladding is a welding operation in which material, usually in powder form, is added to the molten pool and solidifies to produce a surface that has beneficial wear or anti-corrosion properties on top of more easily machinable or less expensive substrate. For example, this process has been applied to automotive valve seats (Matsuyama *et al.*, 2000). When applied to large areas, the process is alternatively called laser hardfacing. The process can also be used to build up worn components by adding the same material as the substrate; in this case, the terminology 'laser weld repair' is appropriate. One of the main areas to which this process has been applied, since the mid-1980s, is in the repair of the tips of the blades of gas turbine engines (Hayes, 1997; Krause, 2001). The added material solidifies epitaxial on the underlaying material, allowing the properties of directionally solidified blades to be maintained in the repair process. The process can be used to reverse machine (that is, to add material instead of machining it away) parts that have been subjected to machining damage, inadvertent damage, or high wear. General Electric Aircraft Engines has applied this process to rebuild turbine spools and disks (Mehta *et al.*, 1984) resulting in considerable cost savings.

Because the laser process involves a low heat input, solidification rates are rapid, leaving little time for segregation of alloy constituents. Under certain processing conditions, a single phase microstructure can be produced via the laser weld repair process, with beneficial surface properties (Hyatt and Magee, 1994). Laser weld repair of nickel aluminium bronze was found to be considerably harder (260 to 377 HV, depending on heat input, versus 265 HV) than similar repairs done by pulsed gas metal arc welding, and have an approximate 30% improvement in resistance to cavitation erosion (Hyatt and Majumdar, 2000). The laser weld repaired material had a factor of 5 improvement in resistance to cavitation erosion compared to the cast base metal and about a factor of 20 decrease in corrosion current (Hyatt *et al.*, 1998). Benefits of the rapid solidification inherent in the laser process have been observed in other alloy systems.

The laser heat input inherent in the laser weld repair process allows one layer to be deposited on another layer, resulting in the building up of three-dimensional structures (Milewski *et al.*, 1998). This allows direct computer aided design (CAD)-to-part manufacture of metallic components, similar to the stereolithography processes that build polymeric components. The process seems to have been developed simultaneously at multiple locations, some targeting general industrial product development, some locations targeting fabricating of complex parts for defense applications. The technology spun off into development of tool and dies, which are made of steels that are challenging to machine. The laser process allowed fabrication of molds with imbedded cooling channels more optimally located than could be achieved by conventional machining processes. The imbedded cooling channels allowed speeding up the injection molding and extraction process, resulting in significant savings to manufacturers. The laser deposition process has allowed large titanium alloy components to be built up for airframes; the traditional manufacturing process required extensive machining of the components from large blocks of the alloy. The laser process achieved a greater utilization of the relatively expensive titanium.

4.10 Safety in laser beam welding

Experts in laser safety divide the potential hazards into two categories, beam hazards and non-beam hazards. The non-beam hazards include factors such as the glare of the welding process, which may contain significant ultraviolet radiation, and fumes from welding. For the most part, the non-beam hazards are similar to those encountered in other welding processes and are discussed in Chapter 10, Occupational health and safety. One additional non-beam hazard that may not be present in other welding processes is the high voltage used for carbon dioxide and flash lamp pumped Nd:YAG lasers. All industrial lasers should be packaged so that operators and engineers cannot possibly have access to the high voltage under normal operating conditions. It is essential that repair technicians be trained to understand high voltage.

Beam hazards is a term used to describe possible eye and skin damage due to contact with a laser beam. All industrial laser processing equipment should be inside a safety enclosure so that welding machine operators cannot under any circumstances come into contact with a beam, hence beam hazards are negligible. In the USA and certain other jurisdictions, this is called a 'Class I' enclosure. Electrical interlocks are located so that if doors to the enclosures are opened, for example to load or unload a part to be welded, laser action is inhibited. It is essential that operators and other staff be trained not to bypass the interlocks.

The wavelength of the output of carbon dioxide lasers is not transmitted through ordinary optical materials such as glass or plexiglass, hence safety enclosures for carbon dioxide lasers can be made of materials that allow the operator and spectators to view the welding process. The wavelengths of the output of Nd:YAG and diode lasers are however transmitted through normal window material, so operators and spectators could potentially come into contact with laser radiation that is reflected from the weldment. Consequently, Class I enclosures for these laser systems normally are made from sheet metal, with either a closed-circuit television camera to ensure the beam is aligned on the seam to be welded, and to allow observation of the weld, or special glass windows through which the beam is not transmitted, or both.

When it is necessary for service personnel to operate the laser without the enclosure, for example for aligning the mirrors in the beam path to ensure the laser beam is centered in the beam path, special precautions must be undertaken. These precautions include the use of wrap-around safety glasses that do not transmit the laser wavelength and safety curtains to ensure that nobody other than the service personnel can encounter the beam. Any laser facility with lasers that are not Class I should have a Laser Safety Officer to evaluate the safety of the laser installation, educate staff that are in contact with lasers, and ensure that adequate administrative and engineering controls are installed to limit the possibility of accidents.

4.11 Future trends

4.11.1 Fiber lasers

Fiber lasers and fiber laser amplifiers were originally developed for the telecommunications industry. As the power output capability increased, fiber lasers were packaged for industrial applications and are finding uses in cutting, welding and drilling. Just as neodymium ions can be doped into

YAG rods and slabs for applications in industrial Nd:YAG lasers, these and other ions can be doped into silica or other substrates that can be drawn into long thin fibers. The active element doped into the fiber can be ytterbium, in which case the wavelength of the output is very close to the wavelength of the Nd:YAG laser. Alternatively, the active element could be erbium, giving a wavelength of 1545 nm. The surface layer of the fiber, called cladding, is doped to give it a higher index of refraction so that it acts like a mirror; the laser radiation is reflected off the walls of the fiber but can be transmitted down the core of the fiber in a nearly loss-less fashion. The core of the fiber is typically 50 µm in diameter; the entire fiber including the cladding is 125 µm in diameter. Outside the cabinet in which the beam is generated, the fiber lies within an armored sheath which provides protection during handling.

The beam transmitted from the fiber is usually of very high quality and can be focused to a small spot size. Fiber lasers can be either continuous or repetitively pulsed. The 'wall plug' efficiency of fiber lasers (ratio of electrical power in to optical power out) is 20% or higher and is the highest of all industrial lasers. Fiber lasers are commercially available at average powers up to the kilowatt range; these high power lasers are made by coupling together the output of many smaller lasers. The fiber laser, because of its exceptionally fine focusing powers, promises to find applications in laser welding. At the time of writing, however, it is too early to evaluate the results.

4.11.2 Combined welding

Combined welding, also referred to as hybrid welding, laser-assisted arc welding, or arc-assisted laser welding, uses the energy input from a laser source as well as the energy input from a gas metal arc torch, a gas tungsten arc torch, or a plasma arc torch. The process was first investigated prior to 1980 but has found commercial applications in manufacture of automobiles and ships in the early part of the twenty-first century. It has been used with both carbon dioxide and Nd:YAG lasers. This process is discussed in more detail in Chapter 6.

Since the combined process makes more efficient use of the relatively expensive laser power, and since the addition of filler metal allows gap tolerances to be relaxed and provides a method of controlling the metallurgy of the weld metal, increasing use of the combined process is expected in the future.

4.11.3 Diode lasers

In passing through the junction between two regions of a semiconductor that have different dopants, an electron loses energy and in some cases can emit this energy in the form of a coherent laser beam. Such laser diodes have a high divergence because of their small size. They can, however, be packaged together in such a way that the output of individual diodes adds to produce a kilowatt level, with specialized optics in the package to control the divergence.

4.11.4 Beam scanning

In most cases, laser welding requires moving a laser focusing head, complete with gas shielding nozzles, over the part to be welded in a controlled manner with a relatively precise standoff distance. This is accomplished through computer numerically controlled workstations which control the movement of the beam over or around the part, or the movement of the part over or under the stationary beam.

In recent years, laser welding systems utilizing galvanometric scanners have been developed by a number of suppliers. Two fast scanners deflect the beam in orthogonal directions; the beam passes through focusing optics and is focused onto the part. A relatively small angular deflection of one of the scanning mirrors results in a relatively large deflection over the part being welded. One commercially available system uses a 1.25 m focal length mirror and can weld over an area of 0.61 m by 1.22 m. It produces 5 spot welds per second with a weld nugget of 3 to 4 mm in diameter. In this case, the target market is the automotive industry. In addition to the speed of welding, another advantage is the long standoff between the scanning and focusing head and the workpiece, allowing parts to be rapidly loaded and unloaded. This technology requires a laser of very good quality, to produce enough intensity to make the weld using the long focal length lens. Since it is impractical to apply a shielding gas over such large weldments at high speed, the process can only be used on material that can be welded in air. Moreover, unless multiple scanning systems are used or the part is manipulated, the spot welds can be applied from one side only, by line-of-sight from the focusing mirror.

4.11.5 Welding with preheat

Often, welding processes including laser welding require that a part be preheated to produce the desired metallurgical properties of the welded materials, or to prevent cracking. For example, laser weld repair of valves was accomplished in a four station work system where the parts were loaded or unloaded in station one, subjected to preheat with a gas torch in position two, laser welded in position three, cooled down in position four, and then rotated back to position one where the parts were unloaded and replaced. The timing was controlled by the length of the welding time in position three with the heat input in position two adjusted so that an adequate preheat was obtained. In other cases, inductive heating has been used to provide preheating for the laser welding process.

A related area in which preheat is beneficial is that where a significant amount of filler metal is required. Laser welding is not an efficient process when the laser energy is used to melt a large amount of filler metal. When it is desirable to weld with a filler metal, there are potential advantages to using a 'hot wire' feed, where the wire is externally resistively (and hence inexpensively) heated to a temperature close to its melting point prior to the wire being injected into the weld pool created by the laser. In spite of its potential advantages, this process has not been extensively investigated or utilized up to the present time.

4.11.6 Multiple beam welding

There have been a number of instances where the use of two laser beams simultaneously has increased the laser beam welding capability. The laser welding had been limited at high speeds due to an instability of the molten pool known as the 'humping' instability (Albright and Chiang, 1988) which was observed as a regular bulging and constriction of the weld face. Banas (1991) developed the process of twin-spot welding using a spherical mirror hinged along its center so that relative angular deflections of the two halves produced two focal regions with an adjustable gap between them. With the laser power split between the two spots aligned along the direction of motion, laser welding of steel could be carried out at higher speeds without the occurrence of the humping instability. For example, in welding 1.5 mm stainless steel, the onset of the humping instability was shifted from 15 m/min to almost 30 m/min using the twin-spot technique. The method was initially developed for tube welding but has been adopted for tailor blank welding and other high speed welding applications.

In welding tee-joints in thick section steel, for example in assembling stiffeners for deck plates in shipbuilding, one approach is to laser weld with the beam at a glancing incidence along one side of the joint, then repeat the weld from the other side of the joint to form a completely penetrating joint. It is found, however, that the first weld produces some distortion of the metal, due to the triangular shaped partial penetration fusion zone. It might be thought that the second weld produces similar but opposite distortion, resulting in straightening, but in fact some residual distortion remains. This distortion problem was eliminated when the weld was executed from both sides simultaneously, using a split beam. A feature of the two-sided welding is the linking together of the two keyholes, resulting in a fusion zone that maximally overlaps the seam.

The disadvantage of the two-beam welding techniques is the need for special optics to split the beam, a challenge due to the high powers involved. Moreover, careful alignment of the beam on the beam-splitting mirror is required to ensure approximately equal power in the two beams. In the case

108   New developments in advanced welding

of split-beam welding of sheet metal, some work has been done using a focusing optic that produces an elongated beam, rather than using two distinct spots.

4.12 Sources of further information and advice

4.12.1 Professional bodies

There exists a number of professional bodies in various countries around the world whose members can provide information and advice, as summarized in Table 4.1. The predominant body is probably the Laser Institute of America which in spite of the 'America' in the name is an international organization. For example, at their 2002 annual conference, the *International Conference on Applications of Lasers and ElectroOptics*, or ICALEO, only about half the attendees were from the USA; there were 101 attendees from 14 European countries, and 43 attendees from six Asian countries, in spite of limitations on travel due to political events and epidemics.

Table 4.1 Professional bodies/industrial organizations

Country	Name of organization	Contact information
USA	Laser Institute of America	12424 Research Parkway, Ste 125, Orlando, FL, 32826, USA
United Kingdom	Association of Industrial Laser Users	Oxford House, 100 Ock Street, Abingdon, Oxfordshire, OX14 5DH, UK
Former Soviet Union	Laser Association (LAS)	Russia, 117485, PB 27 Moscow
Japan	Japan Welding Engineering Society, Laser Materials Processing Committee	Kanda-Sakuma-cho 1–11, Chiyoda-ku, Tokyo, Japan 101-0025
China	Laser Processing Committee of China Optical Society	c/o Minlin Zhong, Secretary-General Department of Mechanical Engineering, Tsinghua University, Beijing 100084, PR China

4.12.2 Research groups

There are several research groups with considerable skill in laser processing. Table 4.2 gives a representative sample.

Table 4.2 Research organizations

Country	Name of organization	Contact information
USA	Edison Welding Institute	1250 Arthur E Adams Drive, Columbus, OH, 43221, USA
USA	Applied Research Laboratory, Penn State University	PO Box 30, State College, PA, 16804, USA
United Kingdom	The Welding Institute	Granta Park, Great Abington, Cambridge, CB1 6AL, UK
India	Centre for Advanced Technology (CAT), at Bhabha Atomic Research Centre	P.O. CAT, Indore 452 013, M.P. India
Ukraine	Laser Technology Research Institute, Kiev Polytechnic Institute	01056 Kiev, Ukraine, PR. Peremohy, 37
Canada	Integrated Manufacturing Technology Institute, National Research Council of Canada	800 Collip Circle, London, Ontario, N6G 4X8, Canada
Japan	Welding Research Institute, Osaka University	11-1 Mihogaoka, Iaraki, Osaka 867-0047, Japan
Australia	Commonwealth Scientific and Industrial Research Organization (CSIRO) Manufacturing and Infrastructure Technology	PO Box 4, Woodville, South Australia 5011, Australia
Finland	Laser Processing Center, Lappeenranta University of Technology	Tuotantokatu 2, FIN-53850 Lappeenranta, Finland
Germany	Fraunhofer Institute, e.g. Institut Werkstoff-und Strahltechnik	Winterbergstrasse 28, 01277 Dresden, Germany

4.12.3 Laser beam welding standards

For the first two decades of laser welding, there were no widely recognized standards regulating the process. Many companies that depended on the process developed their own procedure specification and qualification procedures, or used those previously employed for electron beam welding. In the last dozen years there has been a flurry of international activity. The European standard committee and the ISO organization have proposed and approved a variety of standards, largely relating to equipment. In 2002, the standard *Specification and Qualification of Welding Procedures for Metallic Materials – Welding Procedure Specification – Part 4: Laser Beam Welding*, ISO/FDIS 15609-4:2002 was circulated for approval. Other relevant documents include ISO/DIS 15614-11, *Specification and Approval of Welding Procedures for Metallic Materials – Welding Procedure Test, Part II, Electron and Laser Beam Welding*. Further information and copies of the specifications are available from the ISO secretariat, at International Organization for Standardization, DIN Burggrafenstr 6, 10787 Berlin, Germany.

The American Welding Society has compiled an American national standard ANSI/AWS C7.2 *Recommended Practices for Laser Beam Welding Cutting and Drilling*, based upon input from a committee of laser suppliers, users and researchers. It is expected that a *Process Specification and Operator Qualification for Laser Beam Welding* to be released in 2006. These standards are available from the American Welding Society at 550 NW LeJeune Road, Miami, FL 33126, USA.

The need for and the existence of these standards and specifications indicates that laser welding is now a mature process, accepted for industrial practices.

4.13 References

Albright C.E. and Chiang S. (1988), 'High speed welding instabilities', *Journal Laser Applications*, 1988, 18–24

Anon (1995), 'Nippon Steel to invest $70,000,000', *The Laser's Edge*, **1**(3), 6

Banas C.M. (1972), 'Laser welding developments', *Proceedings CEGB International Conference on Welding Research Related to Power Plants*, University of Southampton, England, Sept 17–21 (1972), London, Central Electricity Generating Board

Banas C.M. (1986), 'High power laser welding', in Belforte D. and Levitt M. (eds), *Industrial Laser Annual Handbook*, Tulsa, Pennwell Books

Banas C (1991), 'Tech Update: Tube Welding', *The Laser's Edge*, December 1991, 2–3

Blake A. and Mazumder J. (1982), 'Control of composition during laser welding of aluminum magnesium alloy using a plasma suppression technique', *Proceedings International Conference on Applications of Lasers and Electro Optics (ICALEO)*, The Laser Institute of America, Orlando FL. Sept 20–33, 1982, Boston, Mass

Boechat A.A.P., Su D. and Jones J.D.C. (1993), 'Dependence of output near-field beam profile on launching conditions in graded-index fibers used in delivery system for Nd:YAG lasers', *Applied Optics,* **32**, 291–7

Bolin S.R. (1976), 'Limited penetration laser welding applications', *Australian Welding Journal*, January/February 1976 23–8

Bolin S.R. (1983), 'Nd:YAG laser applications survey', in Bass M. (ed.), *Laser Materials Processing*, Amsterdam, North-Holland, 407

Brushwood J. (1984), *Assessment of Materials-Processing Lasers*, EPRI EM-3465. Palo Alto, Electric Power Research Institute

DeMaria A.J. (1976), 'Review of high power CO_2 lasers', in Bekefi G (ed.), *Principles of Laser Plasmas,* New York, J Wiley and Sons

Dowden J., Davis M. and Kapadia P. (1983), 'Some aspects of the fluid dynamics of laser welding', *Journal Fluid Mechanics,* **126**, 123–46

Fuerschbach P.W. and Hinkley D.A. (1997), 'Pulsed Nd:YAG laser welding of cardiac pacemaker batteries with reduced heat input', *Welding Journal: Welding Research Supplement*, March 1997, 103S–109S

Graham M.P. and Weckman D.C. (1995), 'A comparison of rotating-wire-and rotating-pinhole-type laser beam analyzers when used to measure pulsed Nd: YAG laser beams', *Measurement Science and Technology*, **6**, 1492–9

Hayes R.H. (1997), 'An innovative technology for the gas turbine component repair industry', *World Aviation Gas Turbine Engine Overhaul and Repair Conference*

Hyatt C.V. and Magee K.H. (1994), 'Laser surface melting and cladding of nickel aluminum bronzes' in *Proceedings of Advanced Methods of Joining New Materials II*, Miami, The American Welding Society, 111–26

Hyatt C.V. and Majumdar A. (2000), 'Effect of microstructure on the erosion corrosion and cavitation erosion behavior of laser clad nickel aluminum bronzes', *Proceedings 12th International Federation of Heat Treatment and Surface Engineering Conference*, Melbourne, Australia Melbourne, Institute of Materials Engineering, 119–21

Hyatt C.V., Magee K.H. and Betancourt T. (1998), 'The effect of heat input on the microstructure and properties of nickel aluminum bronze laser clad with a consumable of composition Cu-9.0Al-4.6Ni-3.9Fe-1.2Mn', *Metallurgical and Materials Transactions A*, **29A** 1677–90

Kawai Y., Alhara M., Ishii K., Tabuchi M. and Sasaki H. (1984), 'Development of laser welder for strip processing line', *Kawasaki Steel Technical Report*, **10**, 39–46

Koechner W. (1976), *Solid State Laser Engineering*, New York, Springer Verlag

Krause S. (2001), 'An advanced repair technique: laser powder buildup welding', *Sulzer Technical Review*, **4**, 4–6

Locke E.V., Hoag E.D. and Hella R.A. (1972), 'Deep penetration welding with high power CO_2 lasers', *IEEE Journal of Quantum Electronics*, QE-8, 132–5

MacCallum D.O., Fuerschbach P.W., Milewski J.O. and Piltch M. (2000), 'Beam characterization of high power fiber delivered Nd:YAG lasers', at the *American Welding Society Convention, Chicago IL*, April 25–27

Matsuyama H., Kano J., Shibata K. and Ninomiya R. (2000), Process and materials development for laser cladding valve seats on aluminum engine heads', *Powertrain International*, **3**(2), 40–8

Mehta P., Cooper E.B. and Otten R. (1984), 'Reverse machining via CO_2 laser', in Mazumder J. (ed.), *Proceedings of the Materials Processing Symposium, International Conference on Applications of Lasers and ElectroOptics*, Orlando, Laser Institute of America, Orlando, FL, 168–76

Metzbower E.A. (1981), 'Laser welding', *Naval Engineers Journal*, August 1981, 49–58

Milewski J.O., Lewis G.K., Thoma D.J., Keel G.I., Nemec R.B. and Reiner T.R.A. (1998), 'Directed light fabrication of a solid metal hemisphere using 5-axis powder deposition', *Journal of Materials Processing Technology*, **75**, 165–72

Notenboom G. (1984), 'Laser spot welding in the electronics industry, in Crafer R., (ed.), *Laser Welding, Cutting, and Surface Treatment*, Abington, UK, The Welding Institute

O'Brian R.L. (1991), *Welding Handbook*, 8th ed., Volume 2, *Welding Processes*, Miami, American Welding Society

Offenberger A.A., Kerr R.D. and Smy P.R. (1972), 'Plasma diagnostics using CO_2 laser absorption and interferometry', *Journal Applied Physics*, **43** 574–7

Ono M., Shiozaki T., Ohmura M., Nagahama H. and Kohno K. (1996), 'High power laser application for steel industry' in *Proceedings 6th European Conference on Laser Treatment of Materials*, September, Stuttgart

Patel C.K.N. (1969), 'High power carbon dioxide lasers', in Schawlow A.L. (ed.), *Lasers and Light, Readings from Scientific American*, San Francisco, W.H. Freeman and Company

Swift-Hook D.T. and Gick A.E.F. (1973), 'Penetration welding with lasers', *Welding Journal: Welding Research Supplement*, 488S–492S

Zacharia T., David S.A., Vitek J.M. and DebRoy T. (1990), 'Modeling the effect of surface active elements on weld pool fluid flow, heat transfer, and geometry', in David S.A. and Vitek T.M. (eds), *Proceedings of the 2nd International Conference on Trends in Welding Research,* Gatlinburg TN, May 14–19, 1989, Materials Park Ohio USA, ASM International

5
Nd:YAG laser welding

M. NAEEM, GSI Group, UK and
M. BRANDT, Swinburne University of Technology, Australia

5.1 Introduction

Nd:YAG lasers have been commercially available for over 30 years. The Nd:YAG (neodynium doped yttrium aluminium garnet) crystals in these lasers can be pumped either using white light flashlamps or, more efficiently, using laser diodes. The Nd:YAG laser is one of the most versatile laser sources used in materials processing. The relative robustness and compactness of the laser and the possibility for the 1.06 µm light it produces to be transmitted to the workpiece via silica optical fibres are two features which contribute to its success. Nd:YAG lasers when first commercialised operated mainly in pulsed mode, where the high peak powers which can be generated were found useful in applications such as drilling, cutting and marking. These pulsed lasers can also be utilised for welding a range of materials. More recently, high power (up to 10 kW), continuous wave (CW) Nd:YAG lasers have become available.

Because of the wide range of applied power and power densities available from Nd:YAG lasers, different welding methods are possible. If the laser is in pulsed mode and if the surface temperature is below the boiling point, heat transport is predominantly by conduction and a conduction limited weld is produced. If the applied power is higher, for a given speed, boiling begins in the weld pool and a deep penetration weld can be formed. Figure 5.1[1] shows the power densities, in W/cm^2, for various laser processes. Pulsed Nd:YAG lasers provide higher power density than any other available source.

5.2 Laser output characteristics

Nd:YAG lasers are characterised by the their power output over time i.e. pulsed, CW and supermodulated (a feature exclusive to GSI Lumonics' new JK continuous wave products).[2]

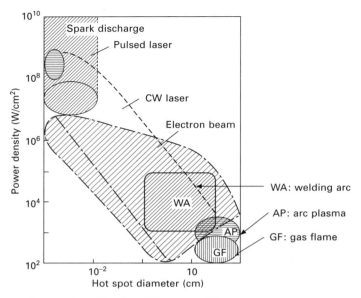

5.1 Power densities for different welding processes.

5.2.1 Pulsed output power

Pulsed Nd:YAG lasers employ a power supply designed for delivering high peak powers during the laser pulse and do not have CW capability. Pulsing implies that the laser's active medium is excited by a very quick response stimulus. This allows the laser to transmit a burst of energy for a brief length of time (generally in terms of milliseconds). Peak pulse powers for pulsing Nd:YAG lasers can reach values of over 30 times greater than the maximum average power levels. This allows low-to-medium power lasers to achieve enough energy to reach vaporisation temperatures for most materials.

The basic laser pulse from the pulsed laser is a rectangular pulse with an initial overshoot spike as shown in Fig. 5.2. Often the single sector standard pulse is quite adequate when welding standard ferrous alloys without a coating or carrying out standard pulsed YAG cutting applications. However, with most welding reflective or dissimilar materials, pulse shaping has a measurable effect on the quality and consistency.[3] Most lasers are rated by their CW output, but pulsed lasers have pulsed energy, peak power, pulse width and frequency terminology that must be understood.

Pulse energy

The volume of the melt puddle for each pulse is determined by the pulse energy. There is a minimum pulse energy required for weld penetration to a

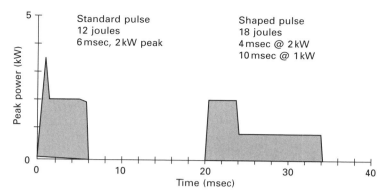

5.2 Basic laser pulse with an initial spike and shaped pulse.

certain weld depth for a given material. Energy per pulse in joules (E) is related to the average power in watts (P) and the pulse frequency (f) by the following:

$$E(\text{J}) = P(\text{W})/f(\text{Hz}) \tag{5.1}$$

Peak power

Height of the pulse is the peak power as shown in Fig. 5.2 and this peak power breaks down reflectivity and overcomes thermal diffusivity. The high peak power is required for precious metal welding and for a range of aluminium alloys. The peak power (P_p) can be calculated by the following:

$$P_p(\text{kW}) = E(\text{J})/t(\text{s}) \tag{5.2}$$

Pulse frequency and overlap

During pulsed Nd:YAG welding, seam welds are produced by a series of spot welds. The pulsing rate of the laser results in faster or slower seam welding as the rate is increased or decreased. To produce hermetic welds, pulse rate (f), spot diameter (d) and the weld speed (v) have to be matched to produce the required percent overlap (%OL). Generally speaking typical values for hermetic welds are between 70 and 80 %OL and for non-hermetic welds between 50 and 60 % OL. The percent overlap can be calculated by the following:

$$\%\text{OL} = 100[(d - v/f)/d] \tag{5.3}$$

Figure 5.3 shows relationship between welding speed and frequency for three different percent overlaps.

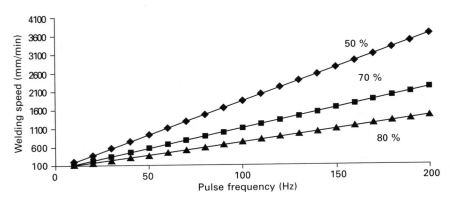

5.3 Welding speed vs. repetition rate for different percent overlaps (spot size 0.60 mm).

5.2.2 CW output power

Continuous wave (CW) Nd:YAG lasers produce a constant output power without interruption and can usually be varied from about 10% to 100% of their mean power rating. GSI Lumonics' proprietary super modulation (SM) technique involves storing some energy in the laser's power supply during the off time of the laser or when the laser is operating below its rated power, and then quickly sending this stored energy to the laser's lamps for extra bursts of peak power. The peak power attainable can be as much as 200% of the laser's mean power. In this way, a supermodulated laser can operate at CW just like any other but can also be directed to produce a square wave, sine wave, or other repetitive output with peak powers above the mean power rating while also producing full mean power. For example, in a 50% duty cycle square wave output, the laser will produce 200% of the CW rating during the laser's pulse 'on-time', thereby producing an average output equal to the laser's full rating. Typical laser outputs are shown in Fig. 5.4. The momentary increase in peak power provided by supermodulation produces some exceptional results during welding (see later sections).

5.2.3 Choosing pulsed or CW

- Minimum heat input: Pulsed Nd:YAG is the choice. If components have metallurgical constraints on heat input or there are heat-sensitive components nearby such as glass-to-metal seals or o-rings, the pulsed YAG can be set up to achieve the required processing rate at a heat input low enough not to damage the components.
- Speed: CW Nd:YAG is the best choice. Whether cutting or welding, by processing the component with a CW beam there is no need to overlap

Nd:YAG laser welding 117

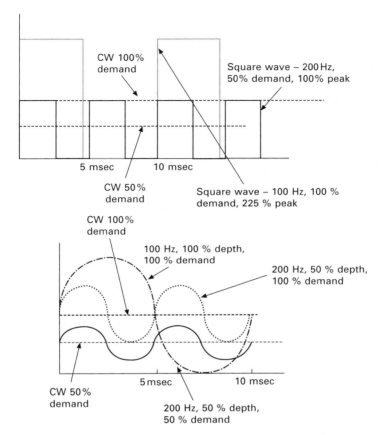

5.4 CW output vs. square and sine wave supermodulation waveforms.

pulses or to re-establish the keyhole. Simply adjust power and speed along with the focus spot size to achieve the desired penetration.
- Welding reflective materials: Usually pulsed Nd:YAGs. For copper and precious metals the pulsed Nd:YAG has the peak power to break down the reflectivity. Only very high average power CW Nd:YAGs can process these materials.
- Heat treating/cladding: Usually CW Nd:YAGs. Average power tends to be the limit to speed, case depth, or remelt thickness. Pulsed Nd:YAGs can do the job but their lower average power ratings rule them out except for small devices.
- Spot welding: Usually pulsed Nd:YAGs. By setting the pulse parameters correctly, the pulsed laser is the fastest and most repeatable spot welder. Only if large diameter nuggets are required would a CW laser be considered.

118 New developments in advanced welding

- Low penetration welding: CW laser will weld very quickly and produce parts with high throughput. Pulsed lasers might have sufficient speed also and have the benefit of dealing with material changes or spot welding requirements.
- Welding crack-sensitive alloys: CW Nd:YAG is the best choice unless there are other constraints such as heat input. The slower cooling rate of the CW laser usually reduces cracking tendencies. This is true of steel alloys containing sulphur, phosphorus, lead, and/or selenium. Also for welding mild steel to stainless steel or steels with poor Cr: Ni equivalent ratios.

5.3 The Nd:YAG laser

The Nd:YAG laser is a solid-state laser, usually in the shape of a rod, operating at 1.06 µm.[4] The active species are neodymium ions present in small concentrations in the YAG crystal. Both continuous wave and pulsed laser outputs can be obtained at an overall efficiency in the 3 to 5% range. This laser is used in industry because of its efficiency, output power and reliability compared to other solid-state lasers. The crystal is grown using the Czochralski crystal growing technique[5] which involves slowly raising a seed Nd:YAG crystal from the molten crystal constituents to extract an Nd:YAG boule.

A single boule typically yields several laser rods. The concentration of Nd ions in the boule is carefully controlled and is no greater than about 1.1%. Increasing the Nd doping further in order to increase the laser power produces unacceptable strain in the crystal and leads to a dramatic reduction in laser power.

Laser rods are typically 6 mm in diameter and 100 mm in length with the largest commercial size rods being 10 mm in diameter and 200 mm in length. Because of the small size of the crystal Nd:YAG lasers tend to be much more compact than are CO_2 lasers. Illustrated in Fig. 5.5 are the main components of a single-rod Nd:YAG laser.

Laser action is achieved by exciting the crystal optically by lamps placed in close proximity to it. The lamps have an emission spectrum, which overlaps the absorption bands of the Nd:YAG crystal at 700 nm and 800 nm. In order to couple the maximum amount of lamplight into the rod and to extract the maximum laser power from it, the rod and the lamp are enclosed in specially designed and manufactured cavities. The two most common pump cavity configurations are elliptical and close coupled. In the case of elliptical cross-sections, the rod and the lamp are placed along the two foci, and in the case of close-coupled cavities, the rod and the lamp are placed close together at the axis. The inside surface of the cavity is normally coated with gold in order to maximise the coupling of lamplight into the rod. Some laser manufacturers also produce ceramic cavities which allow more uniform

Nd:YAG laser welding

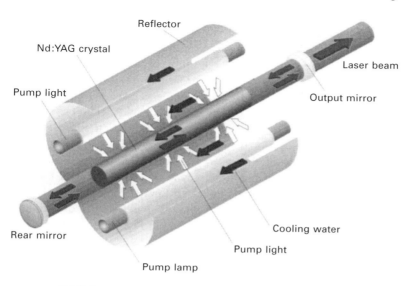

5.5 Schematic of an Nd:YAG laser (courtesy Rofin–Sinar).

pumping of the rod but at the expense of lower efficiency (approximately 5% lower) compared with that of the gold-coated cavities.

For continuous operation, krypton arc lamps are most widely used while for pulsed operation high-pressure xenon and krypton flashlamps are used. Lamp lifetime dominates the service requirement of modern Nd:YAG lasers. For arc lamps, the lifetime ranges between 400 and 1000 h while for pulsed lasers it is about 20 to 30 million pulses depending on operating conditions.

Only a fraction of the emitted spectrum is absorbed by the laser crystal and the rest of the emitted light is dissipated as heat in the cavity and it must be removed for efficient laser operation. This is usually achieved by flowing deionised water around the rod and lamp in a closed loop cooling system. The loop is coupled to a heat exchanger for efficient heat removal.

To increase the laser power above 500 to 650 W, typically obtained from a single rod, requires an increase in the laser volume. However, increasing the rod volume has fundamental limitations. Heat generated within the rod causes large thermal gradients which lead to variations in the refractive index, lowering beam quality, as well as large mechanical stresses, which can cause rod fracture. To obtain higher laser powers involves the use of multiple laser rods. The rods are arranged in series and located either entirely within the resonator or with some being placed outside the resonator to act as amplifiers. These configurations are discussed and described in more detail by Rofin–Sinar.[6] There are now on the market several systems all giving in excess of 2 kW of laser power with the highest power commercial device producing 5 kW from eight cavities.[7]

5.3.1 Diode and diode pumped Nd:YAG laser

While lamps have been an integral part of the Nd:YAG laser technology to date and will remain so for the foreseeable future because of their relatively low cost, another technology is now emerging for high power laser applications both as a pumping source for Nd:YAG lasers and as a laser source in its own right.[8,9] This technology is the high power laser diode and its main advantage lies in having a very narrow spectral output compared with that of the lamp which is matched to the absorption bands of the Nd:YAG laser thus increasing considerably the efficiency of the laser system. Diode-pumped Nd:YAG lasers have much better beam quality because of lower induced thermal stresses, are more compact, require smaller chillers and have much longer lifetimes in comparison with those of the lamps. A schematic of a diode-pumped Nd:YAG laser arrangement is illustrated in Fig. 5.6 for a single rod system. Rofin–Sinar is now offering commercially a 4.4 kW diode-pumped Nd:YAG laser with guaranteed 15 000 h diode operation.

As well as pumping, Nd:YAG rods diode lasers are now being offered as standalone units for high power surfacing and welding applications. The advantages of diode lasers include high efficiency (up to 50%), which leads to lower operating costs and small size and so makes integration into existing production systems relatively simple. Units with output powers in the 3 kW to 5 kW range are now available commercially. The main disadvantage of laser diodes is their cost, which is about US$150–200 per watt of diode power, and the lack of field lifetime data. However, regardless of these

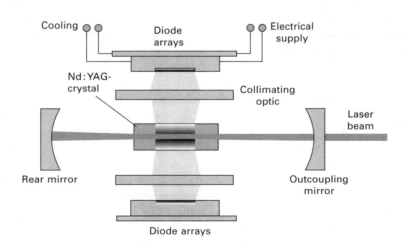

5.6 Schematic of a single rod diode-pumped Nd:YAG lasers (courtesy Rofin–Sinar).

factors, the future for high power diode lasers in manufacturing is bright and their price will decrease as the production volume increases.

5.4 The laser as a machining tool

As a machining tool the laser alone is very ineffectual. It must be used in conjunction with several different items of optical and mechanical equipment, which need to be integrated into a functional unit, in order to be able to process materials. Illustrated in Fig. 5.7 is a typical laser machining system. It comprises three key elements:

- laser source
- beam delivery and processing optics
- multi-axis, numerically controlled motion system.

Of the many laser sources that have been discovered over the years, the CO_2 and the Nd:YAG lasers now dominate in industrial environments. The beam delivery systems for the two laser sources differ in detail. In the case of CO_2 lasers, the system comprises optical elements located within rigid tubes which transport the beam from the laser and focus it onto the workpiece surface. The system may include a number of optical components such as telescopes for diverging the beam, beam splitters for the sharing of power between different processing optics, polarisers for producing circularly polarised beams, energy-share modules for delivering the beam simultaneously at several locations as well as mirrors for bending the beam. In the case of Nd:YAG lasers, the beam can be guided both by mirrors and by glass optical fibres making delivery of the laser beam to difficult and tight spaces relatively easy.[10,11] Nd:YAG lasers with output powers of 5 kW transmitted through 0.6 mm diameter fibres are now commercially available.[7] At present there are no commercial fibres for high power CO_2 lasers.

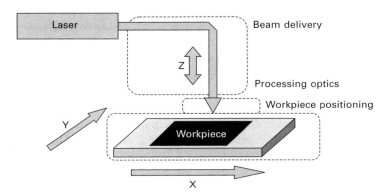

5.7 Components of a laser machining system.

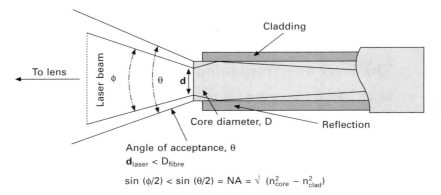

5.8 The coupling of a laser beam into a fibre.

The fibres manually used with industrial Nd:YAG lasers are of the 'step index' design. This means that they have a core with a high refractive index surrounded by a cladding with a lower refractive index (Fig. 5.8). Transmission of light occurs by total internal reflection at the core/cladding interface due to the difference in refractive index between the core and the cladding. A property of these fibres is that the exiting beam has a relatively homogeneous intensity distribution over its diameter. Depending on the power of the laser, the core diameters range in size from 0.2 to 1.0 mm. The fibres are manufactured from high purity fused silica and possess minimal loss at the laser wavelength. Optical losses of the order of 8% per fibre occur in fibres which have no coatings on the ends. As this loss can cause problems at high laser powers, companies such as Rofin–Sinar have developed special coated quartz blanks at the ends of the fibre, which reduce this loss to less than 2%.[12]

To launch a laser beam into a fibre, so that it experiences a minimal transmission loss as it propagates along the fibre, requires that the diameter of the focused spot on the fibre face is smaller than or equal to the fibre core diameter. The focused spot size in commercial systems is normally 80% to 90% of the core diameter, which allows for easier adjustment of the fibre and for any variation in the spot diameter due to laser parameters.[10] In addition, the divergence of the input laser beam must be less than the acceptance angle of the fibre defined by its numerical aperture, NA.

The laser beam exiting the fibre diverges, so to generate a high power density on the workpiece, an optical system is used to recollimate and focus it onto a workpiece. The diameter of the focused spot is determined by the magnification M of the optical system, where M = diameter of focus spot/diameter of fibre core. Typical magnification ratios are 0.5, 1 and 2, which generate spot sizes of 0.3 mm, 0.6 mm and 1.2 mm with a 0.6 mm diameter fibre.

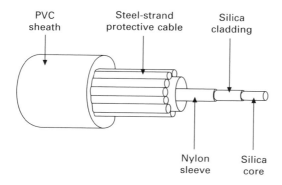

5.9 Schematic diagram of a fibre-optic cable construction.

Illustrated in Fig. 5.9 is a schematic diagram of a fibre-optic cable construction.[11] In addition to the core and cladding layer, most optic fibres used with high power lasers now include continuity detection, which senses if an accidental burn-through has occurred and turns the laser off. For some applications, the fibre is contained within a steel-armoured cable to prevent any mechanical damage. Typical fibre lengths are from 5 to 20 m with a number of reports[14] indicating that lengths in excess of 100 m can be used effectively.

Raw laser beams normally do not have sufficient intensity to cause melting or vaporisation of materials. To increase the intensity in order to process materials, laser beams are focused using both lenses and mirrors. Lenses are generally used with laser powers up to several kW; beyond this catastrophic damage can occur, particularly in the case of CO_2 lasers. At higher powers, mirrors are used because of their high power handling capability. The characteristics of a focused laser beam are shown in Fig. 5.10. The key parameters are the focused spot diameter, d, and the depth of focus, L, defined as the distance over which the focal spot size changes by +/–5%. The focused spot diameter affects the maximum irradiance that can be achieved while the depth of focus influences the process working range.

For circular beams the focused spot diameter, d, is proportional to f/D, where f is lens focal length and D beam diameter at the lens, whereas the depth of field, L, is proportional to f^2/D^2. As can be seen, the two quantities work in opposition. To obtain the smallest spot diameter and therefore the highest power density, the focal length should be small. To obtain the greatest depth of field the focal length should be large. So a compromise must be made to ensure that the correct processing conditions are maintained. The precise spot diameter and depth of focus also depend on the mode structure of the beam as well as on the optical aberrations of lenses and mirrors such as spherical aberration, astigmatism and thermally induced distortion. All

124   New developments in advanced welding

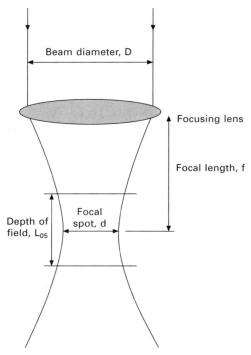

5.10 Focusing lens characteristics.

these quantities tend to increase the spot diameter or shorten the focal length, which can dramatically affect the process.

The material commonly used for lenses and mirrors for Nd:YAG lasers is borosilicate crown glass, designated BK7.[15] BK7 has excellent optical and thermal properties and is relatively cheap. Lenses are normally coated with antireflection coatings to minimise reflection losses at the laser wavelength. The optical materials used with CO_2 lasers are somewhat diverse depending on the laser power and operating conditions.[16] Both reflective and transmissive optics are used. Focusing lenses can be made from gallium arsenide (GaAs), potassium chloride (KCl) and zinc selenide (ZnSe). ZnSe is now the most commonly used lens material because of its very low absorption and high transmission in the visible part of the spectrum, allowing red laser diode lasers to be used to align the invisible CO_2 beam on the workpiece. The lenses are coated with antireflection materials to minimise the nearly 17% loss of the incident laser power that would otherwise occur at each surface. A limitation with ZnSe lenses is that some absorption of laser radiation does occur and as the laser power increases, the absorbed laser radiation becomes significant causing the lens to heat up. This heating changes the imaging properties of the lens, most notably shortening its focal length. The focal length change can be severe enough to cause process capability loss. Because of this problem ZnSe lenses tend to be used with laser powers up to about 3 kW and metal focusing optics above this level. Recently developed air cooled doublet lenses, however, show potential for overcoming the thermal lensing effect and extending the operating laser power levels up to 20 kW.[13] Focusing mirrors tend to be made from copper because it is highly reflective at the laser wavelength and can withstand high energy densities, above 100 kW/cm^2, without sustaining thermal damage. Because copper is soft, the mirrors usually have a coating to protect the surface and are water cooled to minimise thermal distortion. Metal focusing optics is normally used for welding and surfacing applications.

In order to control the motion between the laser beam and the workpiece, a number of approaches can be used, the choice of which depend mainly upon the laser and the workpiece to be processed. For processing flat sheet, material and tubes, two-dimensional systems are employed. These systems involve moving the workpiece using rotational and translational stages while the beam remains stationary, moving the laser beam while the workpiece is stationary or a combination of the two. The systems involving stationary beams allow a high degree of repeatability and accuracy at high processing speeds (up to 20 m/min). The main problem with this method is that the process area is limited because of the large overhang of the machining head required with large sheets. In addition, as one of the axes is positioned on top of the other, the weight of the workpiece becomes an issue. The system is therefore mainly used for precision processing of relatively small parts.

In robot systems, there are also a number of possibilities: moving the laser, the beam or the workpiece. Robot systems with moving optical systems are further divided into systems with interior or external beam guidance. Robotic applications involving Nd:YAG lasers are now becoming increasingly common because of the ease of manipulating the processing head as well as the ability to transmit the beam over long distances.

5.5 Laser welding with Nd:YAG lasers

Laser welding represents a new process being applied to an old industrial technique. It is a fusion welding process requiring no filler material, where parts are joined by melting the interface between them and allowing it to solidify.[17] The process is not just an alternative to conventional welding processes, but rather it offers the engineers and designers greater flexibility in selecting components from materials, which are difficult or impossible to weld conventionally. Much literature now exists on laser welding and the reader may refer to Duley[18] for the most up-to-date discussion and presentation on the theoretical and practical aspects of the process.

The main features of laser welding which make it an attractive alternative compared to conventional processes are:

- Precise narrow and deep welds can be produced with high metallurgical quality (weld bead less than 1 mm and penetration up to 50 mm).[19]
- A small heat-affected zone reduces metallurgical damage and also allows welds to be made close to heat-sensitive components.
- The low heat input into the material obviates the need for complex jigging and allows distortion-free welding of thick to thin sections.
- High process speed–welding speeds in excess of 10 m/min can be achieved with materials of thickness about 1 mm.
- Flexibility allows one laser to be shared among a number of workstations.
- Post-weld treatment is not normally required.
- Welds can be performed in difficult geometries and dissimilar material thicknesses.

Laser welding is performed by one of two mechanisms illustrated in Fig. 5.11. In conduction welding, overlapping spots from a pulsed laser or from the beam of a continuous laser are absorbed by the surface of the material and the volume below the surface is heated by thermal conduction producing a semi-circular cross-section. This type of welding is usually confined to materials up to 2 mm thick. When the laser power exceeds 1 kW and power density exceeds 10^6 W/cm^2 deep penetration welding is achieved. At this intensity level the rapid removal of metal by vaporisation from the surface leads to the formation of a small keyhole into the workpiece. The keyhole grows in depth because of increased coupling of radiation into the workpiece, through multiple reflections of the laser beam off the keyhole walls, and

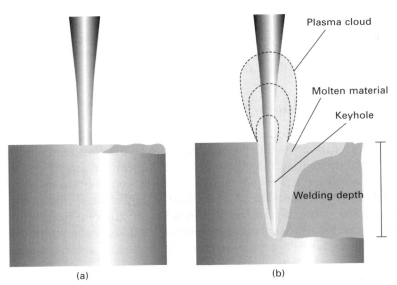

5.11 Principle of (a) conduction welding and (b) keyhole welding.

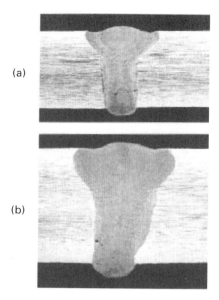

5.12 Transverse sections of butt joint in stainless steel with a 3 kW Nd:YAG laser: (a) 5.7 mm thick @ 2.8 m/min; (b) 8 mm thick @ 0.12 m/min.

material vaporisation. The balance between the hydrostatic forces of the liquid metal surrounding it governs its existence and the pressure of vaporised and ionised material or plasma within it. A typical macrograph of a keyhole weld is shown in Fig. 5.12.

Sometimes plasma is ejected from the keyhole, forming a cloud above the workpiece. This plasma cloud can have a deleterious effect on the welding process because it can shield the workpiece from the laser beam leading to wider and shallower welds. To overcome the problem a shielding gas is normally employed both to suppress plasma formation and to protect the weld from oxidation. Gas flow rates are in the range 10 to 40 l/min depending on the laser power. Helium shielding gas is used when welding with high power CO_2 lasers because of its high ionisation potential which inhibits plasma formation. Oxygen-free nitrogen is also an effective plasma suppressor but can cause embrittlement in some steels. Carbon dioxide gas can be used with pulsed lasers but is avoided with continuous lasers because it assists plasma formation. For the production of long welds in easily oxidised materials where an additional trailing gas cover is required to prevent oxidation, argon shielding gas can be used.

Laser welding, like other processes, requires the control of a number of operating parameters including power, mode, shielding gas and travel speed. There are many compilations of data that indicate the typical welding speeds

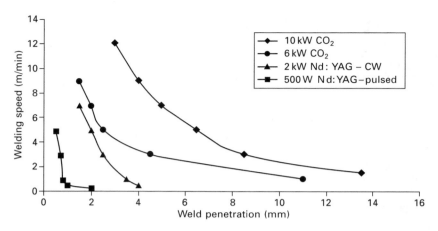

5.13 Representative weld speeds as a function of penetration in mild steel for CO_2 and Nd:YAG lasers.

and weld bead cross-sections, which may be expected.[18] An example of welding speed as a function of penetration for a range of lasers and output powers is illustrated in Fig. 5.13. As a rule of thumb[17] 1 kW of CO_2 laser power at a welding speed of 1 m/min and focusing optic with an f number in the range 6 to 9 gives approximately 1.5 mm penetration in steel.

When considering laser welding the design of the joint and possibly the fabrication method of the whole product should be reassessed in order to gain the maximum advantage from the process. Laser welding requires high tolerances in gap control and joint positioning (Fig. 5.14). Using filler material can widen the tolerance field but, in practice, this is not very common because of the associated reduction in weld speed. In addition to the suitable preparation and alignment of the joint faces, they should be free from contaminants such as grease, paint, dirt and oxide scales. Residue from chemical degreasing and cleaning agents should be carefully removed since weld spatter and porosity can result if these substances are present on the workpiece.

In principle, a laser can also weld any material that can be joined by conventional processes. Illustrated in Table 5.1 is the weldability of metal pairs. In the welding of dissimilar metals, good solid solubility is essential for sound weld properties. This is achieved only with metals having compatible melting temperature ranges. If the melting temperature of one material is near the vaporisation temperature of the other, poor weldability is obtained and often involves the formation of brittle intermetallics.

Nd:YAG laser welding

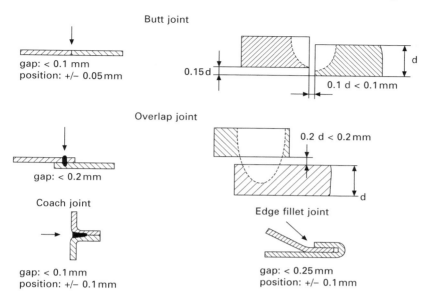

5.14 Maximum gap and positioning tolerances for different laser-welded joints.

Table 5.1 Weldability of metal pairs

	Al	Ag	Au	Cu	Pd	Ni	Pt	Fe	Be	Ti	Cr	Mo	Te	W
Al	◆													
Ag	○	○												
Au	○	◆	◆											
Cu	○	○	◆	◆										
Pd		◆	◆	◆										
Ni	○		◆	◆	◆	◆								
Pt		○	◆	◆	◆	◆	◆							
Fe			○	○	●	●	●	◆						
Be			○	○	○	○			○					
Ti	○	○	○	○	○	○	○	○		◆				
Cr		○			●	●	◆	◆			●			
Mo						○	◆	●		◆	◆			
Te					●	●	○	○		◆		◆		
W					○	○	●	○		○	◆	◆	●	◆

◆ Very good   ● Good   ○ Sufficient

5.6 Nd:YAG laser welding tips: process development

5.6.1 Check solid solubility of the major constituents

This is not a final factor in determining feasibility but a quick check is worthwhile. Almost all materials are alloys, i.e. combinations of many metals, so looking at each constituent is not practical or worthwhile. If certain metals are required, try them at the prototype level.

5.6.2 Are these alloys being welded by another process?

If the metals to be laser welded cannot be welded by processes such as tungsten inert gas (TIG), they might not be good candidates for laser welding. Check with a welding engineer or someone with a considerable degree of experience with the alloys in question.

5.6.3 Check any platings and coatings

The platings and coatings include phosphorus in electrodeless nickel plating, galvanising, lead and tin in solder coatings, oxide coatings, carburised surfaces, and paint or anodised coatings. Any coating can cause a problem – phosphorus, lead and tin lead to the formation of brittle intermetallics, oxide coatings cause weak welds and porosity, the high carbon content of carburised surfaces results in brittle steels and organic contaminants from paint cause porosity and other problems. Zinc with its low boiling point can cause metal loss and porosity in galvanised steels. If unsure of the effect of platings and coatings, weld the samples without the coatings to determine their effects. Be careful of variations in coatings and platings from batch to batch.

5.6.4 Is the fit-up acceptable?

Laser welding is usually performed without any filler metals so gaps must be filled with metal from the adjacent area and bridging a gap requires extra laser energy. Generally, a gap no more than 10% of the thickness of the thinnest component is allowed. This can be relaxed for thicker materials greater than 1 or 2 mm but might need to be reduced to 5% or lower for materials less than 0.2 mm. Gap problems show themselves as very concave weld beads or failure to bridge the gap between parts. It is also much more difficult to start a laser weld in a large gap area in comparison with seam welds through a short section of high gap area where bridging can sometimes be maintained.

5.6.5 What is the throughput requirement?

High throughput welding jobs usually require multiple laser sources and/or the very high welding speeds of CW lasers. Fast spot welding is best done with pulsed lasers. Optimise the laser welding process to meet all requirements of strength, distortion, heat input, etc. After the weld process has been developed throughput decisions can be made, such as the fastest laser source, automation and special beam delivery that will not compromise weld and part quality.

5.6.6 Is there a total heat input requirement?

Only pulsed lasers can produce welds using extremely low average power no matter what the penetration. Are there glass to metal seals, plastic, or electronic components with a maximum temperature rating? Lasers have much lower heat input than most other welding technologies so supermodulated CW sources might have sufficiently low heat input.

5.6.7 Is there a beam delivery constraint?

Nd:YAG lasers, both CW and pulsed, can be delivered with fibres or conventional mirror delivery. Long focal length lens requirements can favour mirror-based beam delivery but usually a design can be produced for either beam delivery. Talk with a laser applications engineer if the standard beam delivery systems will not work.

5.6.8 What type of weld joint is required?

Lasers can produce butt, lap, and fillet joints. Know the benefits and drawbacks of each and try to design the weld joint to make the best use of the technologies.

5.6.9 Centreline cracking?

Look for intermetallic forming constituents from platings or free-machining alloys. If there are dissimilar materials, weld each material separately as a bead-on-plate test. The material that cracks in this test is usually the crack contributor. If both materials weld without any problems, look for solid solubility problems or high stress in the weld zone such as that found with fillet welds or poor joint fit-up. Try to have material certificates for each metal in use, each time it is used, especially in the prototype stages; the prototype builder may use incorrect alloys.

5.6.10 HAZ cracking?

Look for high carbon constituents that produce brittle phases at the heat affected zone (HAZ) in iron alloys. Steels such as 1035–1070 or stainless steels such as 430–440 have this problem. Try using a CW laser for high carbon content steels or a pre-heat to reduce post weld cooling. Some super alloys also exhibit this type of cracking due to precipitation hardening in the HAZ. Hastelloy 713C and waspalloy exhibit this type of cracking.

5.6.11 Extensive cracking throughout the weld?

Some alloy combinations will show cracking throughout the weld, both centreline and transverse cracking, as well as cracks that move into the parent metal. This is very common for high-strength steels and for some aluminium alloys. A pre-heat is required for the high-strength steels to eliminate the problem. For aluminium alloys the addition of another aluminium alloy at the weld joint such as 4047 or a 5000 series alloy is required to overcome the cracking. Some 300 series stainless steels generally considered very easy to weld will show this problem and it is due to the very high cooling rate of pulsed laser welding and a slight variation to the nominal alloy combination or chromium and nickel and their 'equivalents'. Moving to a CW laser or pulse shaping will help with this but good material control is the best method.

5.6.12 Pinholes and/or random cracking?

Look for very low melting-point contaminants such as lead, zinc, tin, or other organics such as solvents, plastics, oils and fibres. Some plating contains organic brighteners that contribute the volatile constituents. Re-welding over the affected area can reseal over the holes since many of the volatile components are boiled out of the melt puddle on the first pass and the second pass will seal the unit. Initial part cleanliness and good housekeeping are the best solutions here.

5.6.13 Unexplained results?

Check with a welding engineer or other welding expert. Try other processes to determine if they also show the same effect to try to determine potential causes. Talk with material suppliers concerning best practices with their alloys and any issues with storage, heat treatment, or condition of the materials you have.

5.7 Nd:YAG laser welding of different metals

Interest in welding applications for high-power Nd:YAG lasers is growing as the availability of average power, pulsed and CW models of these lasers increases. These lasers offer processing rates and capabilities that can compete with the industry standard CO_2 laser welder, but have the added benefit of flexible fibre optic beam delivery. Pulsed and CW Nd:YAG laser-welding processes differ in performance characteristics, weld shapes and applications, even though both types of lasers produce energy at the same wavelength, i.e. 1.06 µm. This section will review typical laser welding capabilities of both pulsed and CW lasers in a range of materials.

5.7.1 Steels

The largest market for laser welding is the automotive industry where it is being applied to the welding of thin, typically 0.7 to 3 mm thick, coated and uncoated steels, transmission components and the fabrication of sub-assemblies. Perhaps the most significant laser welding application is laser welding of tailored blanks.[20,21] The process involves welding sections of steel sheet of different thickness together first and then stamping the product to form a part. It offers significant cost and environmental benefits over the traditional approach in which a range of techniques are applied to preformed parts which are then subsequently welded; it is now transforming the approach to car body manufacture. Other significant areas where laser welding is making an impact include the electronics and aerospace industries where the applications include hermetic sealing of electronics packages, joining difficult-to-weld electronic materials and the fabrication of components made from high-performance alloys.

For laser welding of steel sheet, there are two main factors to consider, namely the effect of steel composition and the effect of coating.

Effect of steel type

For low carbon steel sheet, CO_2 and Nd:YAG laser welding will produce welds consistently. Compared with the parent material, the hardness of the welded joint is, in general, increased by a factor of 2.0–2.5. This increased hardness can influence the formability as well as the dynamic mechanical properties (e.g. fatigue/impact) of the welded joint.

More recently, there is a growing tendency within the automotive industry to use high strength steels, such as high strength low alloy (HSLA) or microalloyed (Nb, Ti and/or V), rephosphorised, bake-hardened, dual phase or trip steels, as they allow weight reductions to be achieved. Although little laser processing data on these types of steel is yet available it is considered possible to weld most, but care should be taken in monitoring maximum weld hardness and susceptibility to cracking. Microalloyed steels will produce higher weld hardnesses at the same welding conditions in comparison with cold rolled mild steels. Their higher hardnesses could cause problems in post-weld processing operations or in the dynamic performance of the welded structure and alterations in welding conditions to reduce heat input and cooling rate may be necessary.

Effect of coating type

The effect of coatings on the welding process has been the subject of extensive research. Although a range of coatings can be applied to steel sheet, such as

Al, Zn–Al, Zn–Ni or organic coatings, only zinc-coated steels are considered here; they are most commonly used in the automotive industry.

The presence of zinc in the coating, which boils at 906 °C, can cause blowholes and porosity along the weld seam. This usually occurs if the sheets are clamped tightly together and when the coating thickness on the sheets is in excess of 5 µm. A common solution is to create a gap at the joint interface enabling the Zn vapours to escape, which can be done with a roller adjacent to the weld point, the use of special clamping arrangements or dimpled sheets. The use of proprietary gas mixtures or special welding parameters involving pulsing can also be applied. The use of rollers and specially designed clamping systems, however, seems to be the preferred industrial option for the production of three-dimensional laser welds on steel sheet structures. The dimpled sheets add an extra operation and the successful use of special welding parameters is dependent on the coating type and thickness. More recent studies have also reported some success by using twin beam techniques, because they produce a slightly elongated weld pool and therefore give the Zn vapours more time to escape.

In terms of coating type, the three most common zinc-based coatings used are electrogalvanised, galvannealed and hot dipped galvanised. In general, hot dipped galvanised coatings are thicker and can create more problems with porosity and blowholes in the weld. In addition, variability in the thickness of coating can create difficulties in producing consistent welds. Thickness variation should be controlled to ± 2 µm if possible along the joint length.

A complex coating, for instance a zinc layer underneath a chromium/chromium oxide top layer or a thin organic layer (< 0.1 µm), also places extra demands on the laser welding process. Although these materials can be welded, it is possible that extra porosity is generated in the weld due to degradation of the coating.

For lap joints, the main difficulty is the presence of the coating at the interface between the two sheets. If the weld solidifies rapidly, Zn vapours can be entrapped in the weld and cause porosity. For butt joints, for tailored blanks for instance, the coating does not generally cause significant porosity, but the laser welding process does remove the coating from around the weld, leaving an area that may be susceptible to corrosion. However, the removal of coating from the weld is much localised (< 2 mm from the weld centre) and the surrounding coating can offer galvanic protection.

5.7.2 Aluminium alloys

Aluminium alloys are used in a wide range of industrial applications because of their low density and good structural properties. Laser welding has been identified as a key technology that can offer distinct advantages over conventional joining techniques such as TIG, MIG (tungsten and metal inert gas, respectively), resistance spot welding, mechanical fasteners and adhesive bonding.

The main problems associated with laser welding of aluminium alloys in general are the high surface reflectivity, high thermal conductivity and volatilisation of low boiling point constituents. These and other material-related difficulties can lead to problems with weld and HAZ cracking, degradation in the mechanical properties and inconsistent welding performance. These problems are now largely overcome with the advent of higher average powers, improved beam qualities giving a power density high enough to produce a stable keyhole for welding. At present, both CO_2 and Nd:YAG lasers can be used successfully for welding a vast range of aluminium alloys, with slightly higher welding speeds achievable for Nd:YAG lasers compared with similar power CO_2 lasers, because of shorter wavelength (1.06 μm) and improved coupling.

The greatest recent drive towards use of aluminium-based structures has arisen mainly from the automotive industry (e.g. Audi A8, new Audi A2). The requirement for reduced vehicle weight has led to the development of aluminium frame structures (assembled from cast and profile materials), aluminium alloy sheet assemblies and the use of lightweight cast components. Both the 5xxx (aluminium–magnesium) and 6xxx (aluminium–magnesium–silicon) series aluminium alloys are candidate materials and these alloys can be welded with or without filler wire. For a given power density and spot size, the laser welding speed (Fig. 5.15) for 5xxx series alloys is slightly higher than that for 6xxx series alloys and it is believed that this is caused by magnesium vapours stabilising the keyhole.

Although it is possible to weld most aluminium alloys, some are susceptible to weld metal or HAZ cracking. This is especially the case for the 6xxx series alloys, where cracking has been related to the formation of Mg–Si

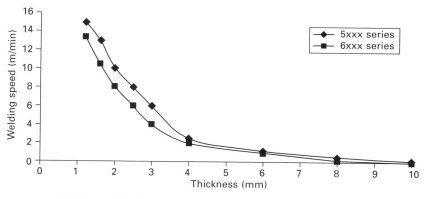

5.15 Material thickness vs. welding speed for aluminium alloys (3.5 kW, 0.45 spot size).

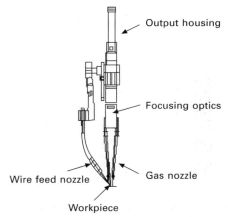

5.16 Wire feed laser-welding set-up.[22]

precipitates. This cracking can be reduced or eliminated by addition of correct filler wire during welding (Fig. 5.16) which reduces the freezing range of the weld metal and minimises the tendency for solidification cracking. The use of filler wire also improves the fitup tolerance and weld profile and can improve the cross-weld strength and elongation to failure value of the joint. Typical welding and wire speeds for 6xxx series alloys are shown in Tables 5.2 and 5.3 respectively. Figure 5.17 shows tensile strength results for different shield gases (6181 aluminium alloy).

Table 5.2 Overlap welds results (average power = 3.50 kW)

Alloy type	Thickness (mm)	Spot size (mm)	Power density (W/cm^2)	Welding speed (m/min)	Wire feed rate (m/min)
6082	2*1.2	0.45	2.20*10^6	5.0	3.0–4.5
6181	2*1.2	0.30	4.95*10^6	6.0	2.5–3.0
6181	2*1.2	0.45	2.20*10^6	5.5	3.0–4.5
6181	2*1.2	0.60	1.24*10^6	4.5	3.0–5.0
6023	2*1.2	0.45	2.20*10^6	5.5	3.0–4.5

*Indicates 2 layers, each 1.2 mm thick

Table 5.3 Welding speeds of aluminium sheet alloys with and without filler wire (average power = 3.50 kW)

Alloy type (thickness)	Filler wire	Spot size (mm)	Welding speed (m/min)	Wire feed rate (m/min)	Gap (mm)
6082 (1.6 mm + 1.6 mm)	–	0.60	10	–	–
6082 (1.6 mm + 2.0 mm)	4043	0.60	6	3	0.30
6023 (1.5 mm + 1.5 mm)	4043	0.60	8	3.5	0.45
6181 (1.6 mm + 1.6 mm)	4043	0.60	7.5	3.5	0.30

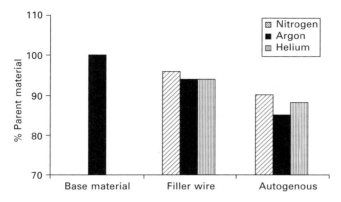

5.17 Tensile strength results for different shield gases, 6181 aluminium alloy.

Although no special surface treatment is required when welding aluminium, care has to be taken to avoid excessive porosity. The predominant cause for porosity is the evolution of hydrogen gas during weld metal solidification. This hydrogen can originate from lubricants, moisture in the atmosphere and surface oxides or from the presence of hydrogen in the parent material. Good quality welds can be achieved for most alloys by cleaning the surfaces prior to welding and adequate inert gas shielding of the weld pool.

Whereas high power CW Nd:YAG is best suited for welding aluminium alloy sheet metal up to 3 mm for automotive applications, pulsed Nd:YAG is better suited for the welding of electronic packages. This is because its pulsing capabilities can deliver the power to the workpiece with minimal heat input. When a designer requires a lightweight, corrosion-resistant, heat-dissipating, robust, and economical package, aluminium is usually the first choice. Aerospace packages for microwave circuits, sensor mounts, or small-ordinance imitators are the most common examples of aluminium components that can be laser welded. Aluminium alloy type 6061-T6 is the material of choice because of its rigidity, ease of machining and economic considerations. However, the material cannot be successfully laser welded to itself, because the partially solidified melt zone cannot withstand the stress of shrinkage upon solidifying and cracks are formed (termed 'soldification cracking' or 'hot cracking'). The solution to this problem is to improve the ductility of the weld metal by using aluminium with a high silicon content such as alloy 4047 (Al 12% Si). This alloy is very ductile as a solid and difficult to machine into small complex shapes. Therefore, 6061 is usually employed as the package component with intricate features, and 4047 is used as a simple lid that is relatively thin (typically less than 1 mm). A 4047 ribbon can be inserted between 6061 components to produce excellent welds, but this requires a very labour intensive step, unless round washers or other simple preform geometries can be employed.

138   New developments in advanced welding

Alloy 2xxx (Al–Cu) and many other popular aluminium alloys are also weldable using 4047 filler metal. So far, there has been no experience indicating that 2xxx can be welded to itself without the use of filler material. The only alloys that can be welded with low heat input and with any filler material are 1000 and 1100 series alloys. These commercially pure alloys have the metallurgical characteristics that enable them to avoid hot cracking, but their poor mechanical and machining properties usually prohibit use in most applications.

5.7.3 Stainless steels

Stainless steels are chosen because of their enhanced corrosion resistance, high temperature oxidation resistance or their strength. The various types of stainless steel are identified and guidance given on welding processes and techniques that can be employed in fabricating stainless steel components without impairing the corrosion, oxidation and mechanical properties of the material or introducing defects into the weld. The unique properties of the stainless steels are derived from the addition of alloying elements, principally chromium and nickel, to steel. Typically, more than 10% chromium is required to produce a stainless iron. The four grades of stainless steel have been classified according to their material properties and welding requirements:

- austenitic
- ferritic
- martensitic
- austenitic–ferritic (duplex and super-duplex).

When laser welding these steels care is required in the selection of gases and gas-shielding arrangements to produce clean, oxide-free welds.

Austenitic stainless steels

These steels are usually referred to as the 300 series and are generally suitable for pulsed and CW laser welding. Slightly higher weld penetration depths or increased weld speeds can be achieved when compared with low carbon steels (Fig. 5.18) due to the lower thermal conductivity of most stainless steel grades. The high speeds of laser welding are also advantageous in reducing susceptibility to corrosion. This corrosion is caused by precipitation of chromium carbides at the grain boundaries and can occur with high heat input welding processes. In addition, laser welding of these grades results in less thermal distortion and residual stresses compared with conventional welding techniques, especially in those steels having 50% greater thermal expansion than have plain carbon steels. The use of free machining grades should be avoided because these steels contain sulphur that can lead to hot

Nd:YAG laser welding 139

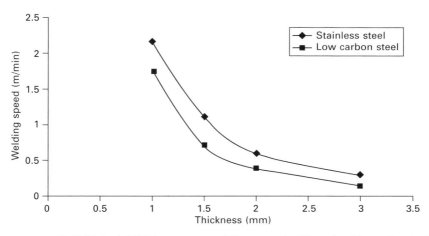

5.18 Material thickness vs. welding speed with pulsed laser (spot size 0.30 mm).

cracking. An excellent example of stainless steel welding with a pulsed Nd:YAG laser is the manufacture of the Gillette sensor razor (Fig. 5.19). Over a hundred lasers with fibre optic beam delivery systems are in operation, producing over 50 000 spot welds and approximately 1900 cartridges each minute.

Austenitic stainless steels are used in applications requiring corrosion resistance and toughness. These steels find wide ranging applications in the oil and gas, transport, chemical, and power generation industries and are particularly useful in high temperature environments. There are a number of potential benefits that result from using high power Nd:YAG laser welding of stainless steels, including productivity increases. The low heat input of the laser welding process reduces the width of the HAZ, thus reducing the

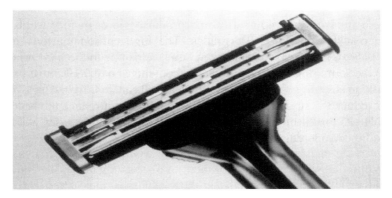

5.19 Gillette sensor razor head, twin stainless steel blades spot-welded with GSI Lumonics pulsed laser.

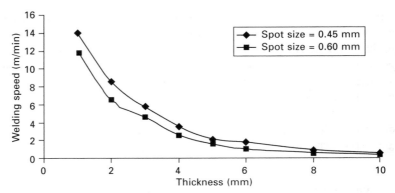

5.20 Material thickness vs. welding speed for 304 stainless steel (3.50 kW, CW).

region that may be susceptible to pitting corrosion. A graph of welding speed against material thickness is shown in Fig. 5.20. The tensile tests on the samples produced an average tensile strength of 98% of the parent material values.

Ferritic stainless steels

These 400 series steels do not possess the good all-round weldability of the austenitic grades. Laser welding of the ferritic grades in some cases impairs joint toughness and corrosion resistance. The reduction in toughness is due in part to the formation of coarse grains in the HAZ and to martensite formation which occurs in the higher carbon grades. The heat-affected zone may have a higher hardness due to the fast cooling rate.

Martensitic stainless steels

These steels are the 400 series and produce poorer quality welds than do either austenitic or ferritic grades. The high carbon martensitic grades (> 0.15%C) can cause problems in laser welds due to the hard brittle welds and the formation of HAZs. If carbon contents above 0.1% must be welded, use of an austenitic stainless steel filler material can improve the weld toughness and reduce the susceptibility to cracking but cannot reduce the brittleness in the HAZ. Pre-heating or tempering at 650–750°C after laser welding may also be considered.

Duplex stainless steel

The introduction of duplex stainless steel for tube and pipe work has created a number of difficulties for the more conventional arc welding processes in terms of achieving the desired phase balance and mechanical/corrosion performance. For tubes or pipework, the manipulation capabilities of fibre optic beam delivery associated with Nd:YAG laser technology and the possibility of remote welding makes high power CW Nd:YAG laser potentially attractive for welding thin-walled duplex stainless pipes. The steel composition, laser parameters and type of gas shielding can influence phase distribution in the weld metal. The low heat input associated with Nd:YAG laser welding can reduce the proportion of austenitic material present in the weld metal and HAZ, which may impair the corrosion properties of the joint.

Titanium alloys

The high strength, low weight and outstanding corrosion resistance possessed by titanium and its alloys has led to a wide and diversified range of successful applications in chemical plant, power generation, oil and gas extraction, medical and especially aerospace industries. The common problem linking all of these applications is how best to weld titanium parts together or to other materials. High power CW Nd:YAG laser welding is one technique that is finding increasing application for titanium alloys. The process, which offers low distortion and good productivity, is potentially more flexible than either TIG or electron beam for automated welding and the application is not restricted by a requirement to evacuate the joint region. Furthermore, laser beams can be directed, enabling a wide range of component configurations to be joined using different welding positions. The welds are usually neat in appearance and have low distortion when compared with their arc-welded counterparts. The fusion zone width and the grain growth can usually be controlled according to laser power at the workpiece and the welding speed used. Although these alloys can be welded without difficulty using lasers, special attention must be given to the joint cleanliness and the gas shielding. Titanium alloys are highly sensitive to oxidation and to interstitial embrittlement through the presence of oxygen, hydrogen, nitrogen and carbon. Laser welding, as arc welding, requires the use of an inert shield gas to provide protection against oxidation and atmospheric contamination. The most frequently used cover gases are helium and argon. Figure 5.21 shows typical welding speeds for Ti-6Al-4V alloy with argon shield gas. Full penetration welds can be produced up to 12 mm thick. The results show that with optimum laser and processing parameters it is possible to produce porosity and crack free welds in both alloys.

While the high power CW Nd:YAG process is mainly used for welding thick sections, pulsed Nd:YAG lasers are largely used for small components that require very little heat input. One such component is the heart pacemaker as shown in Fig. 5.22. The pacemaker is made of titanium alloy package fabricated from two disk-like halves after pacemaking electronics have been

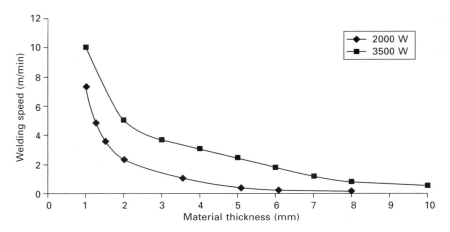

5.21 Welding speed vs. material thickness for Ti-6Al-4V alloy.

5.22 Heart pacemaker welded by pulsed Nd:YAG laser.

sandwiched between them. The sealing weld has to provide a very high quality hermetic seal to prevent body fluids entering the package and causing the pacemaker to fail. The weld also has to take place within 1mm of some of the circuitry and batteries, which should not be subjected to temperatures above 50 °C.

Structural steels

Conventional Nd:YAG laser welding applications have been high speed or precision welding of thin-section materials. However, the introduction of Nd:YAG lasers with 4–5 kW of power opened up many new opportunities for the welding of thicker sections in industries that had not previously considered the process viable. Such applications can be found in shipbuilding, off-road vehicles, power generation and petrochemical industries. For these industries, distortion due to welding is being increasingly recognised as a major cost in fabrication. This realisation led to at least three shipyards introducing high power CO_2 laser welding in an attempt to reduce distortion substantially and improve overall fabrication accuracy. High power CW Nd:YAG laser welding with the benefit of fibre optic beam delivery offers even more potential benefits for thick section welding and a programme has been carried out to asses the potential for the welding of structural steel.

Interest lies mainly in butt welds and T-joints in various linear and circular forms depending on the application. Owing to the limitation at present of typically 3.5–4 kW at the workpiece, the welding speeds for single pass welding of butt joints in 10 mm thick steel are not very high (typically 0.3 m/min). For T-joints, which can be welded from each side, higher speeds are possible.

The main difference between laser welding at pulsed laser power (600 W) and CW laser power (4 kW) is the greater amount of laser-induced plasma/plume emanating from the deep penetration keyhole in the former. This plasma/plume can reduce weld penetration and cause instabilities in the vapour-filled keyhole at the centre of the weld pool, resulting in coarse porosity particularly for materials > 4 mm thick. In order to produce defect-free welds it is essential to suppress the plume/plasma during welding. This usually is done with a helium shield gas ejected through a nozzle.

5.7.4 Nickel-based alloys

These alloys can be laser welded either by pulsed or CW Nd:YAG lasers in a similar manner to that carried out on stainless steels. Comprehensive gas shielding is needed and welding speeds may need to be modified to avoid the occurrence of solidification and HAZ cracking for selective alloys.

5.7.5 Copper-based alloys

These alloys have high reflectivity and thermal conductivity which restricts the penetration capability of laser welds to 1–2 mm. Pulsed Nd:YAG lasers are suitable because of their high peak powers. Laser welding of brass can also suffer from porosity due to vaporisation of zinc.

5.8 Control of Nd:YAG laser welding

Both CO_2 and Nd:YAG laser welding is carried out on those products where a high confidence level in the weld quality is necessary. Some of examples already mentioned are the welding of razors blades, heart pacemakers and automotive parts. The aspects that contribute to producing welds of respectable quality can be grouped under the following headings:

- materials
- joint design
- welding conditions
- in-process monitoring } measurement of quality
- joint tracking

5.8.1 Materials

Variations in materials used can have a significant effect on weld quality. Some relevant variations are the following.

Surface quality

The laser welding process is mainly unaffected by variations in surface quality of the material unless the changes are sufficient to prevent the coupling of the laser beam. This commonly occurs with highly polished surfaces that increase the threshold power density for achieving a keyhole type process. Highly oxidised surfaces can produce porosity during welding.

Coating thickness

Variation in the coating thickness can alter the welding performance. The weld joints are characterised by blowholes or porosity.

Proximity of sealants

If sealants or adhesives are present on the joint line that is to be laser welded, disruption of the weld bead and excessive porosity will occur.

Pressing quality

The most common variation likely to be seen for the three-dimensional structures will be the quality of the pressed components. If the gaps between the joints to be welded are too big and cannot be compensated by the clamping operation, inconsistent weld quality will result.

5.8.2 Joint design

The effect of joint fit-up has already been explained. Fit-ups which leave gaps of > 10% sheet thickness will for butt, lap, hem or edge joints result in a weld undercut which will adversely affect weld properties and performance. Factors affecting joint configurations for laser welding are shown in Table 5.4.

5.8.3 Welding conditions

Focus position

Optimum focus position is dependent on weld joint geometry and weld strength requirements. The optimum focus position is typically that which yields the maximum weld penetration for the butt joint and weld width interface for the lap joint configurations. In general, the tolerance to focus position for laser welding of sheets is ± 0.5–2 mm for Nd:YAG lasers, with the focus being on the top workpiece surface for the focal lengths 80–200 mm. When the focus positions outside these tolerances are used, the weld will show:

- Reduced penetration, if the focus position is above the workpiece surface;
- Greater tendency for the weld undercut, if the focus position is below the workpiece surface.

If the focus position is moved further into the workpiece, a loss of weld penetration will occur.

Welding speed

The welding speed is the parameter most often adjusted when defining optimum welding conditions. This takes into account such factors as laser power, laser mode, spot size and power density. Given that all the other parameters are constant, welding speed or weld penetration will increase with:

- increased average power (Fig. 5.23)
- improved beam quality
- small focused spot size (Fig. 5.24).

If the welding speed is too high, the weld is characterised by a loss of weld penetration and cracking, while with low welding speed the weld exhibits excessive drop trough, top bead undercut, a disrupted weld bead and there may be excessive porosity.

Shielding gas

The shielding gas fulfils two main roles, to provide protection against excessive

Table 5.4 Factors affecting joint configuration for laser welding

Factors	Joint configuration					
	Lap	Butt	Hem	Edge	Multi-layer	T-butt
Tolerance to gap between sheets	<10% ST	<10% ST	<10% ST	<10% ST	<10% ST for each layer	<10% ST
Tolerance to beam joint misalignment	>1mm	<0.3–0.5mm	>1mm	<0.3–0.5mm	>1mm	<0.3–0.5mm
Tolerance to beam focus position	±1mm	±1mm	±1mm	±1mm	±1mm	±1mm
Seam tracking requirement	No	Yes	No	Yes	No	Yes
Tolerance to edge preparation	Avoid burrs	<5% ST	Avoid burrs	<5% ST	Avoid burrs	<5% ST
Tolerance to coatings (e.g. zinc)	Low	Medium	Low	Medium	Low	Medium

Note: ST = Sheet thickness; applies to sheet materials up to 6mm thick

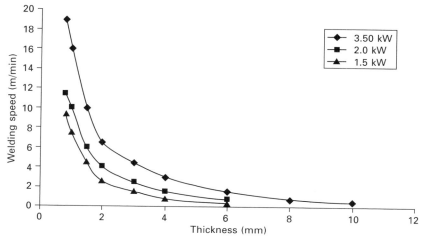

5.23 Welding speed vs. material thickness for C–Mn steel at different CW average powers, spot size = 0.45 mm.

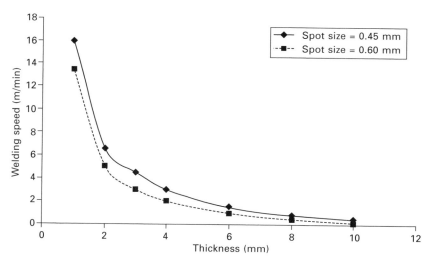

5.24 Welding speed vs. material thickness for C–Mn steel, average power 3.50 kW.

oxidation and to reduce plasma formation. The formation of plasma is more critical to welding when using the CO_2 laser, as there is an interaction involving laser energy and the cloud of ionised gas above the weld, which reduces the penetration. Nd:YAG laser welding does not suffer plasma formation; however, when welding thick sections (> 4 mm) at slow welding speeds, there is a cloud of gas above the weld which can affect the quality of the weld.

The most frequently used cover gas is either helium or argon and typically it is directed centrally at the laser/material interface; if there is an auxiliary tube design, it is directed towards the trailing weld (hot material). Helium is technically the most suitable shielding gas for CO_2 laser welding due to its ability to suppress any plasma formation; in the case of Nd:YAG laser welding, helium gas can also be used for welding stainless steels, aerospace alloys and a range of aluminium alloys. However, due to its low mass, flow rates that provide effective protection from the atmosphere are high, especially for open, three-dimensional components. This factor, coupled with the high cost of helium, makes the use of other cheaper gases attractive.

For whichever type of shielding gas and delivery used, a too low gas flow is characterised by a heavily oxidised weld surface while too high a gas flow causes excessive weld undercut and disrupted weld bead. In most cases underbead shielding is not required for welding at speeds > 1 m/min. However, for stainless steels, nickel alloys, titanium alloys and aluminium alloys the use of underbead shielding is recommended in order to produce an acceptable underbead appearance.

Measurement of quality

With CO_2 or Nd:YAG laser welding the relationship between process parameters and the weld quality is complex. In addition to problems caused by changes in the material composition and surface conditions, alignment errors can be significant, especially when welding large structures. Such errors can affect the weld quality and one approach to detect defect welds is to use in-process monitoring or seam tracking systems; this allows errors to be recognised as they occur.

5.8.4 In-process monitoring techniques for laser welding

The use of optical energy for welding, in the form of a laser beam, offers a number of opportunities for sensing single defects in the process. Thus information is obtained that reflects the processes occurring and the quality of those processes. Figure 5.25 is a diagram indicating the signals that are available from the laser process for in-process monitoring.

For Nd:YAG laser welding systems, applications are partly stimulated by the commercial availability of optical fibre beam delivery systems. This facilitates the integration of laser welding with robotised methods and allows operation on three-dimensional workpieces. During the welding operation, it is very important to monitor the weld quality, reduce the quantity of scrap generated and avoid the possibility of weld failure. This is particularly important in such automotive applications as tailored blank welding or body in white welding, where weld quality and productivity are very important.

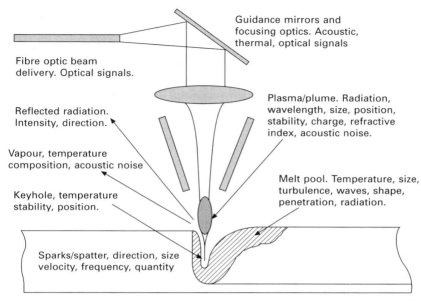

5.25 Opportunities for in-process monitoring of laser processes.[23]

There are a number of potential sensors for weld monitoring in both CO_2 and Nd:YAG laser systems available on the market.[24] These include optical sensors, which detect either the UV/visible light emitted from the plasma/plume that forms above the workpiece during welding, or infrared black body emission from the melt pool. A typical Nd:YAG monitoring set-up system is shown in Fig. 5.26 and was developed by Prometec GmbH. This unit monitors the production process continuously, initiating an alarm for the operator and recording an alarm message when a defect from the normal process occurs. The camera, which is mounted in coaxial alignment to the laser beam, can be used with all current high-power laser types. It monitors and documents several critical process characteristics for example, the melt pool dimensions, penetration depth and gaps, as well as laser parameters. The monitoring system can be adapted to further specific manufacturing needs.

Various monitoring systems are in current use in the automotive industry for tailored blank welding. It has been estimated[25] that over 55 million blanks are welded each year and this figure is growing rapidly. In pursuit of a solution to the problems of weight reduction and over-specification, the concept of tailored welded blank has become an area of particular interest. A tailored welded blank is produced by welding together two or more pieces of sheet to form a single sheet, which is than pressed into a shape. The following parameters play an important part in producing good quality welds

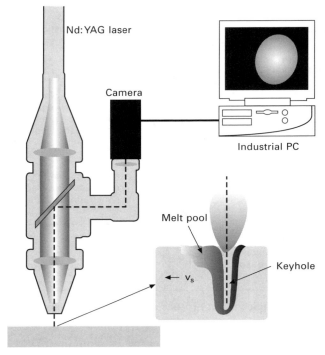

5.26 Integration of a PD 2000 on a typical Nd:YAG laser processing head.

and it is very important to monitor them during welding. Any variation in them will affect the quality of the welds:

- focus position
- joint gap
- laser power
- spot positioning
- indentations (notches) on the edge of the material.

The welding monitor LWM900, commercially available from Jurca Optoelektronik GmbH, processes signals provided by a range of detectors. Each detector measures a different welding process signal that is very or partially independent of the other detected signals. The use of more than one detector is said to improve the correlation between the weld quality and the monitoring results. In production operation, the LWM900 requires several welds, made under optimised production conditions to 'learn' the characteristics of a 'good' weld. The LWM900 (using fuzzy logic) detects process disturbances as signal changes relative to the memorised weld reference. It then calculates the probability that an important weld defect has occurred. In such a case, the calculated probability exceeds a pre-adjusted probability threshold, which is signalled to the machine controller. It is important that the 'normal' weld conditions are stable to ensure a stable baseline. If the nominally 'good weld' conditions vary from part to part, then the baseline will be noisy and it will be difficult to detect deviations from the norm. For research purposes the detected signals processed by the weld monitor can simply be analysed with respect to any given weld imperfection.

As can be seen from the schematic arrangement shown in Fig. 5.27, four detectors were utilised during the experimental trials. The 'plasma detector' sensed the ultraviolet radiation being emitted by the welding plasma/plume. The luminosity of the plasma/plume was then related to weld quality. The

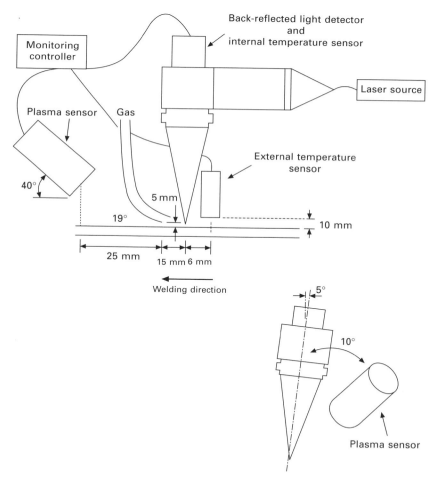

5.27 Schematic arrangement for LMW 900.[26]

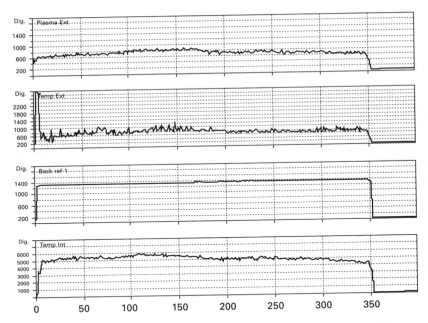

5.28 Typical detector traces produced under optimised welding conditions.

detector for the 'back-reflected laser light' senses radiation with a wavelength of 1.06 μm being reflected from the workpiece during welding with a Nd:YAG laser. It is believed a direct relation exists between the amplitude of this signal and the keyhole geometry; the latter is itself correlated to laser intensity and welding speed. In addition, two 'temperature sensors' were aligned such that they detected the weld pool temperature by measuring infrared radiation. Changes in the thermal capacity of the workpiece during processing should strongly influence these detectors. Some of the output examples from this system will be highlighted.

Figure 5.28 shows detector traces recorded from a good weld produced under optimum welding conditions. A change in the vertical focus position can alter the welding performance. It is possible to detect the change in the focus position and a typical output is shown in Fig. 5.29. Another laser parameter that can affect the welding performance is the laser power. Reduction in the laser power might be caused during the production process through several possibilities such as laser or beam delivery system component failure or a general reduction of power over time as components become degraded. Typical outputs are shown in Fig. 5.30.

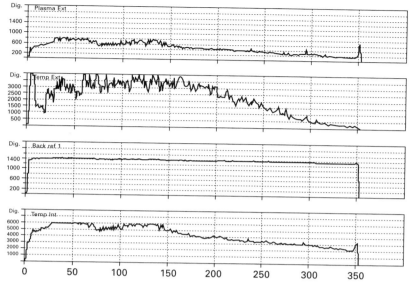

5.29 Detected signals arising from a change in vertical focus position.

5.8.5 A review of joint tracking systems for laser welding

Owing to relatively low tolerance of laser welding to joint misalignment and gap, joint tracking systems offer the potential for improved quality assurance by allowing adjustment for small variations in joint position and fit-up brought about by standard engineering parts tolerances. Joint tracking systems for laser welding applications usually need to locate the joint to between 0.1 mm for 1 mm thick sheet and 0.4 mm for 6 mm thick plate. In addition, the joint tracking systems need to be able to operate at welding speeds of between 1 and 10 m/min for typical plate and sheet applications. The generic differences between laser and arc welding result in the need for different joint tracking system specifications for laser welding. Five types of sensor are generally used for joint tracking.

Tactile sensors

Tactile sensors use a probe or stylus in direct contact with the workpiece; they position the welding torch at the proper location with respect to the joint by either mechanical or electromechanical means. Tactile sensors are of limited use in laser welding, primarily due to the requirement to locate in a mechanical corner and secondly because information is only provided on joint position.

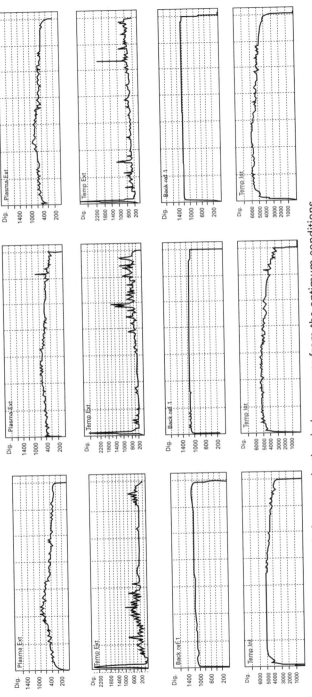

5.30 Detected signals from a reduction in laser power from the optimum conditions.

Vision based sensors

Vision based sensors comprise the greatest percentage of joint tracking systems used today. There are several variants on these systems, but all are centred on the use of a charge-coupled device (CCD) image sensor. At present the sampling rates of the CCDs used in seam tracking equipment are insufficient for some laser welding applications. High-speed cameras are available, but currently at an inadequate size and a substantial cost penalty.

Ultrasonic sensors

Ultrasonic monitoring using a sensor in contact with the workpiece is potentially capable of simultaneous joint tracking in square edge butt joints. However, coupling of the sound waves to the work piece can only be achieved using gel, grease or water. In addition, the sensor needs to be accurately maintained at a constant distance from the weld pool. Airborne ultrasonic sensors are under consideration for future applications but development of suitable air coupled transducers is necessary.

Eddy current sensors

Eddy current sensors use an inductive coil, which sets up a magnetic field in the material and a detector to monitor the field strength in various positions. It is a non-contact device and produces a continuous signal that can be monitored. However, it can only be used with ferrous materials. A number of systems aimed primarily at arc welding applications are currently commercially available. In laser welding, the main concerns involve accuracy, due to local variations in field strength, which can be caused by variable surface quality, and through inherent signal interpretation problems.

Plasma sensors

Plasma sensors use the slight charge in the welding plume to measure the voltage drop from an isolated nozzle to the workpiece. This charge varies with plume intensity and therefore can be correlated with weld quality. It may, therefore, be possible to use this voltage drop to detect variations in joint position. Potential problems, however, lie in the resolution of the technique. The voltages are very low and variations are in the order of millivolts. It is questionable whether there are variations with mistracking and, if there are, whether they are detectable.

5.9 References

1. Ready J.F. and Farson D.F., (eds), *LIA Handbook of Laser Materials Processing*, Magnolia Publishing, 2001
2. www.gsilumonics.com
3. Naeem M., 'The influence of pulse shaping on laser material processing', *9th NOLAMP Conference*, Nordic Conference on Laser Material Processing, Trondheim, Norway, August 2003, 239–49
4. Koechner W., *Solid-state Laser Engineering*, 2nd ed, Springer-Verlag, 1988
5. Bruni F.J. and Johnston G.M., 'Careful system design speeds laser-crystal growth', *Laser Focus World*, May 1994, 205–12
6. Rofin-Sinar technical publication, 2000, Hamburg, Germany
7. www.trumpf.com
8. Bachmann F., 'Introduction to high power diode laser technology', *Proc. 2nd Int. WLT-Conference on Lasers in Manufacturing*, Munich, June 2003. Edited by the German Scientific Laser Society
9. Emmelmann C. and Piening A., 'Diode-pumped solid-state lasers for industrial laser material processing', *Proceedings 32 ISATA*, International Symposium on Automotive Technology and Automation, Prof Dr Dieter Roller (ed.), Vienna, 14–18 June 1999, 359–66
10. Hunter B.V., Leong K.H., Miller C.B., Golden J.F., Glesias R.D. and Laverty P.J., 'Selecting a high-power fiber-optic laser beam delivery system', *Proceedings, ICALEO '96* Laser Institute of America, Detroit, MI, October 1996, Section E 173–82
11. Kugler T., 'Trends in fiber optic beam delivery for materials processing with lasers'. *Optics and Photonics News*, July 1994, 15–18
12. Emmelmann C., *Introduction to Industrial Laser Processing*, Rofin–Sinar Publication, 1998, 34
13. Noaker P.M., 'Welding optics withstand high laser powers', *Laser Focus World*, Feb 1999, 129–31
14. Ishide T., Matsumoto O., Nagura Y. and Nagashima T., 'Optical fibre transmission of 2 kW CW YAG laser and its practical application to welding', *ECO3 Conference on High Power Solid-state Lasers and Applications*, Hague, March 1990, 188
15. Musikant S., *Optical Materials: An Introduction to Selection and Application*, New York, Marcel Dekker, 1985
16. Luxton J.T., 'Optics for materials processing', in Belforte D. and Levitt M., (eds), *Industrial Laser Handbook 1986*, Tulsa OK, Penwell, 1986, 38–48
17. Dawes C., *Laser Welding – A Practical Guide*, Cambridge, Abingdon Publishing, 1992
18. Duley W.W., *Laser Welding*, New York, John Wiley and Sons, 1998
19. Fukuda N., Matsumoto T., Kondo Y., Ohmori A., Inoue K. and Arata, Y., 'Study on high quality welding of thick plates with a 50 kW CO_2 laser processing system'. *Proceedings, ICALEO '97*, Laser Institute of America, San Diego, CA, November 1997, Section E, 11–20
20. Mombo Christian J.C., Lobring V., Prange W. and Frings A., 'Tailored welded blanks: a new alternative in body design', *Industrial Laser Handbook*, Tulka, OK, Penwell Books, 1992, 89
21. Ream S.L., 'Targeting tailored blank welding', *Industrial Laser Solutions*, August, 1999, 7
22. Naeem M., 'Aluminum tailored blank welding with and without wire feed, using

high power continuous wave Nd:YAG laser' IBEC 98, *International Body Engineering Conference & Exposition*, **5**, 247–55
23. Steen W.M., *Laser Material Processing*, 3rd ed., Springer-Verlag, 2003
24. Wlodarczyk G. and Hilton P.A., *Introduction to Laser Weld Monitoring*, TWI CRP Report 629/1998, February 1998
25. Auty T., 'A Simple approach to laser blank welding', *The Industrial Laser User*, **15**, May 1999, 23–5
26. GSI Lumonics Internal Report

6
New developments in laser welding

S. KATAYAMA, Osaka University, Japan

6.1 Introduction

A laser is an outstanding invention of the twentieth century; a variety of lasers have been developed and applied in many industrial fields since Maiman announced laser oscillation in optically pumped ruby crystals in 1960.[1] A focused laser beam is a heat source operated at extremely high power or energy density. It can heat, melt and evaporate any material and consequently produce a deep spot or bead weld at a high speed. It is expected that laser materials processing should play an important role in fundamental and advanced technology in the twenty-first century. Laser welding has received much attention as a promising joining technology because it encompasses high quality, precision, performance, and speed with good flexibility and low deformation or distortion. In addition it allows robotic linkages, reduced man-power, full automation and systematization.

High power lasers (listed in historical order) including CO_2, lamp-pumped YAG, diode (LD), LD-pumped YAG, fiber and disk lasers have been developed as welding heat sources. Combined or hybrid heat sources using two or three lasers with the same or different wavelengths, or a laser and another heat source such as metal inert gas (MIG) and tungsten inert gas (TIG) have been employed to produce better weld beads at high speeds efficiently. Furthermore, a second harmonic Nd:YAG laser with a few milliseconds time period has recently been developed to melt a copper sheet easily (see Chapter 5.) There has been considerable research on laser welding in order to understand the phenomenon or to apply its technology in industry. Relevant titles include 'Laser weldability and welding phenomena of materials such as high strength steels, Zn-coated steels, aluminum alloys, stainless steels, Ni-base superalloys and magnesium alloys', 'Spectroscopic analyses of laser-induced plasma/plume, and elucidation of emission, absorption and scattering properties of plasma' and 'Development of monitoring and adaptive control systems for penetration and weld quality'.

In this chapter, the current state of laser sources for welding is described in terms of their characteristics, merits, power, beam quality and general applications so that the developmental trend of laser apparatuses is understood better. Subsequently, interesting new results, novel interpretation of welding phenomena (including melt flows) and the formation and prevention of welding defects are briefly summarized; the discussion includes the laser welding of stainless steels, zinc-coated steels, aluminum alloys, magnesium alloys, dissimilar metals and plastics. New process developments in remote or scanner laser welding, in-process monitoring and adaptive control during laser welding as well as laser-arc hybrid welding are noted as possible future technologies. These new developments in laser welding are applied in several industries.

6.2 Strengths and limitations of current laser welding technologies

6.2.1 Characteristics, power levels and beam quality of typical lasers for welding

The characteristics, laser media, maximum/normal power levels and merits of typical lasers for welding are summarized in Table 6.1.[2] Lasers have been developed to achieve higher power, higher beam quality and/or higher input-to-output efficiency. The CO_2 laser can provide the highest output power with continuous emission (commercial units up to 45 kW), while lamp- or LD-pumped YAG and fiber lasers can deliver 10 kW class power. Fiber lasers can produce high power and will possibly replace high power CO_2 lasers.

The beam quality is defined as M^2 or BPP (beam parameter product in mm·mrad), as shown in Fig. 6.1.[3] The beam waist of 0.2 mm diameter is shown for various laser types with different BPP in Fig. 6.2.[3] Fiber lasers are now believed to deliver the highest beam quality and the advantages of improved beam quality are summarized in Fig. 6.3.[4] A higher power density can be obtained by a smaller spot size with the same optics, or the same power density can be achieved at lower laser power, leading to reduced cost, as shown in Fig. 6.3(a). The same spot size can be attained at a longer working distance (Fig. 6.3(b)) or with a slim optics of smaller diameter (Fig. 6.3(c)), leading to improved manipulation and enhanced processing operation capability.[4] The correlation of beam quality to laser power for respective lasers is overlaid with the condition regimes for several material processing methods in Fig. 6.4.[2–8] The beam quality of a laser worsens with an increase in power. Deep-penetration or high-speed welding can be generally performed with a high power laser of the 5 kW class, and it is understood that LD-pumped YAG, thin disk, CO_2 and fiber lasers can provide high-quality beams. The quality of high power diode lasers is the worst, although their wall plug efficiency is the highest. The development of higher power CO_2 or YAG lasers is at present fairly static and therefore intensive effort is focused on

Table 6.1 Characteristics, laser media, maximum/normal powers and merits of typical lasers for welding

---

CO_2 laser (wavelength: 10.6 μm; far-infrared ray)
    Laser media   : CO_2–N_2–He mixed gas (gas)
    Average power [CW]   :   45 kW (maximum)
        (Normal) 500 W – 10 kW
    Merit   :   Easier high power (efficiency: 10–20%)

Lamp-pumped YAG laser (wavelength: 1.06 μm; near-infrared ray)
    Laser media   : Nd^{3+}: $Y_3Al_5O_{12}$ garnet (solid)
    Average power [CW]   :   10 kW (cascade type max & fiber-coupling max)
        (Normal) 50 W–4 kW (efficiency: 1–4%)
    Merits   :   Fiber-delivery, and easier handling

Laser Diode (LD) (wavelength: 0.8–0.95 μm; near-infrared ray)
    Laser media   :   InGaAsP, etc. (solid)
    Average power [CW]   :   10 kW (stack type max.), 5 kW (fiber-delivery max.)
    Merits   :   Compact, and high efficiency (20–50%)

LD-pumped solid-state laser (wavelength: about 1 μm; near-infrared ray)
    Laser media   : Nd^{3+} : $Y_3Al_5O_{12}$ garnet (solid), etc.
    Average power [CW]   :   13.5 kW (fiber-coupling max.)
        [PW]   :   6 kW (slab type max.)
    Merits   :   Fiber-delivery, high brightness, and high efficiency (10–20%)

Disk laser (wavelength: 1.03 μm; near-infrared ray)
    Laser media   : Yb^{3+} : YAG or YVO_4 (solid), etc.
    Average power [CW]   :   6 kW (cascade type max.)
    Merits   :   Fiber-delivery, high brightness, high efficiency (10–15%)

Fiber laser (wavelength: 1.07 μm; near-infrared ray)
    Laser media   : Yb^{3+} : SiO_2 (solid), etc.
    Average power [CW]   :   20 kW (fiber-coupling max.)
    Merits   :   Fiber-delivery, high brightness, high efficiency (10–25%)

---

the development of high-power diode, LD-pumped YAG or solid-state, disk and/or fiber lasers with higher beam quality.

The reflectance or reflectivity of near- or far-infrared lasers is high for most metals, but decreases with a decrease in their respective wavelength. Copper vapor lasers ($\lambda = 510$ nm) and the second harmonic Q-switched YAG lasers ($\lambda = 532$ nm), which can melt and evaporate highly reflective metals such as copper, are used for metals drilling. Recently, the apparatus delivering the second harmonic YAG laser of 2 W with 3 ms pulse width has been developed with the objectives of welding copper sheets and so on directly.[9] Such direct welding of copper may well replace soldering. Typical lasers and their features are briefly described in the following paragraphs.

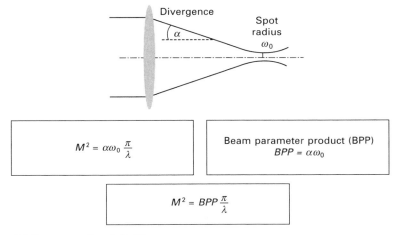

6.1 Beam quality definition.

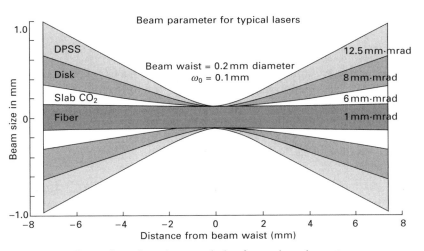

6.2 Beam focusing characteristics for various laser types.

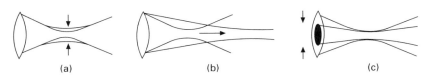

6.3 Effect of improved beam quality on focusing. (a) Smaller focus at constant aperture and focal length, (b) longer working distance at constant aperture and spot diameter, (c) smaller aperture ('slim optics') at constant focal diameter and working distance.

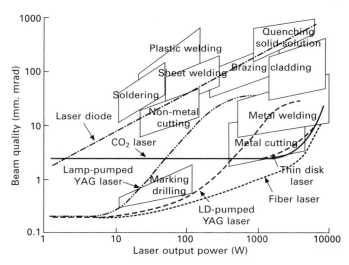

6.4 Beam quality as function of laser output power for respective lasers, overlaid for several laser materials processing.

6.2.2 CO_2 lasers

The highest average output power can be obtained with a CO_2 laser. Lasers in the 2.5 to 7 kW class are normally used in the automobile industry[10–12] and 5 to 45 kW class lasers are utilized in the steel and shipbuilding industries.[13–15] Furthermore, remote or scanner welding technology is noted in some industrial fields, and high quality CO_2 laser systems with output power up to 6 kW levels are now used as heat sources for remote welding of car body components.[16–18]

Cross-sections of laser weld beads in Type 304 steel made at 10 to 40 kW are shown in Fig. 6.5.[19] Deeply penetrated welds of a keyhole type are produced. The penetration depth increases proportionally with an increase in the laser power. Porosity is almost always present in such deeply penetrated welds.[19,20] In Ar or N_2 shielding gas at high powers such as 5 kW and more, Ar or N plasma which blocks laser energy reaching the plate is always or intermittently produced along the laser beam axis over the shot location by the coaxial gas flow torch or by the plasma control nozzle from an oblique angle, respectively.[20] Examples of plasma formation affecting weld penetration during laser welding are schematically illustrated in Fig. 6.6. The tendency of gas plasma to form depends upon the laser power and the material.[19–22] Therefore, in CO_2 laser welding, He shielding or He-mixed gas should be generally used at more than 10 kW to avoid the strong interaction which takes place between a laser beam and Ar plasma in the case of Ar shielding gas.[20]

New developments in laser welding 163

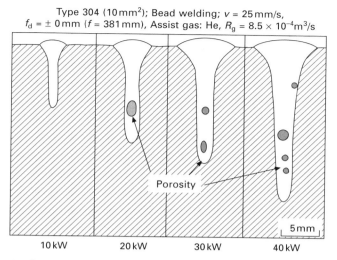

6.5 Cross-sectional weld beads produced in Type 304 steel with a CO_2 laser in He gas at 10 to 40 kW.

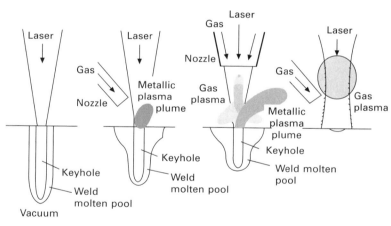

6.6 Schematic illustration showing the effect of plasma formation on laser weld penetration.

Porosity is generally less in full penetration welds than it is in partial ones.[20] In partial penetration welding at low speeds, longer periods of interaction between the laser beam and the keyhole wall and the instability of a deep keyhole during welding are chiefly attributed to bubble generation, leading to porosity formation.[20–22] The formation of bubbles and porosity can be reduced by the selection of proper repetition and width of pulse modulation,[20–22] as shown in Fig. 6.7.[21,22] In this case, the porosity is drastically reduced at

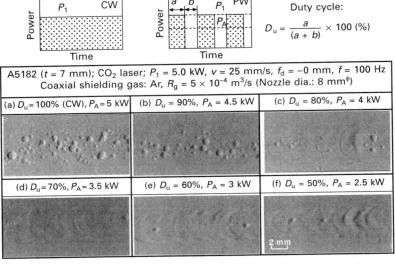

6.7 X-ray inspection results of laser-welded A5182 alloy, showing the effect of pulse repetition on formation of bubbles and porosity.

60 to 70% duty (6 to 7 ms pulse beam irradiation period), because of the suppression of bubble generation.

6.2.3 YAG lasers

A YAG laser can be oscillated in the mode of continuous wave (CW), normal or modulated pulsed wave (PW) or Q-switching. The PW or CW laser beam can be delivered through a GI or SI fiber. In the case of low power or low pulse energy, as indicated in Fig. 6.8,[9] deeper penetration can be obtained by GI fiber than by SI fiber, because the former can provide higher power density under the focused conditions. At high laser powers, there is little difference in the penetration between GI and SI fibers, and SI fibers having higher damage thresholds are chiefly used at CW powers greater than 1 kW.

A normal pulsed YAG laser can be used in spot or seam micro-welding of small parts in the electrical and other industries,[23] and moreover SI-fiber delivered laser apparatus in the 3 to 4.5 kW class is widely used in the automobile industry.[12] In deep-penetration, weld spots and beads produced with pulsed or CW lasers easily cause porosity and therefore, in the case of pulsed laser, saw-like or tailing pulse shape should be utilized to reduce porosity.[24–26] Controlled saw-like pulse shapes can reduce porosity and underfilling in the spot weld by suppressing spattering and adjusting keyhole depth, as shown in Fig. 6.9.[26] In the case of high CW powers, a system using two or three beams was developed for reduction of porosity.[27,28]

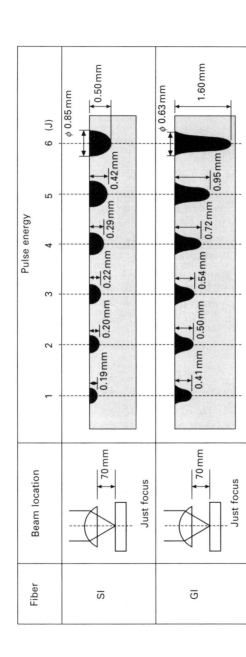

6.8 Comparison of the effect of GI and SI fiber on spot weld penetration.

166   New developments in advanced welding

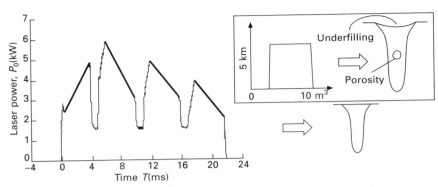

6.9 Special saw-like pulse shape effective for porosity reduction.

YAG lasers of 10 kW class can be obtained by one apparatus[29] or fiber-coupled 3 sets of 3 or 4 kW class lasers.[30] Penetration depths are indicated as a function of welding speed; comparisons between CW and modulated PW are shown in Fig. 6.10.[29] When the average laser output powers are equal, pulse modulation offers greater advantages in the production of deeply penetrated welds owing to its higher power density at lower welding speeds. On the other hand, deeper penetration of a weld bead at higher speeds, i.e. greater than 25 mm/s, as shown in Fig. 6.10, can be achieved in CW mode.

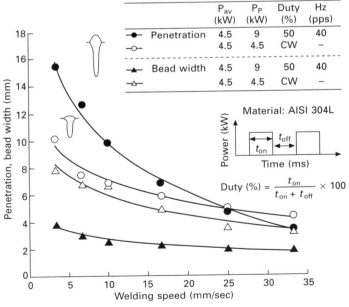

6.10 Effect of CW and PW modulation on penetration depths as a function of welding speed.

6.2.4 Laser diodes (LD)

High power laser diodes (LD) or diode lasers, which can be used directly and/or in fiber-coupled mode, are commercially available in a maximum power output range up to 10 kW and 5 kW, respectively.[4,31–34] Diode lasers of low power levels are suitable for the welding of plastics. High power diode lasers of rectangular beam shape leading to moderate power density (which are generally called bad quality) are directly used to weld thin sheets of aluminum alloys or steels at high speeds. Furthermore, fiber-delivered diode lasers are employed for brazing of Zn-coated steels by using robots together, and some lasers can produce deeply penetrated keyhole-type weld beads in stainless steel at low speeds. It is generally accepted that diode lasers are suitable for the welding of plastics and thin metal sheets as well as the brazing, soldering, quenching, surface melting treatment and cladding of metals.

The development of bright high-power diode lasers is anticipated as compact, highly efficient heat sources for such processes as welding and brazing.[4,34]

6.2.5 LD-pumped solid-state lasers

Commercially available LD-pumped YAG lasers can deliver brighter and higher-quality beams at higher efficiencies than lamp-pumped YAG lasers. LD-pumped YAG lasers of 2.5 to 6 kW power are used in the automobile industry.[4] A 13.5 kW laser system using three sets of 4.5 kW apparatus is in special use.

Rod-type LD-pumped Nd:YAG lasers with output powers of 8 and 10 kW have been realized as a laboratory prototype in Germany and Japan and a slab-type LD-pumped Nd:YAG laser of 6 kW power has been developed by PLM (Precision Laser Machining Consortium) in the USA.[4,35–37] The welding results with the latter slab-type laser are compared with those with a lamp-pumped YAG laser in Fig. 6.11.[37] Weld bead penetration can be extremely deep at low welding speeds. Bead welding was performed with a pulsed laser of focal length 350 mm and about 10 mm depth of focus at a high repetition rate (for example, 400 Hz) with a weaving process because of the narrow beam diameter; thereafter, cosmetic treatment might be required for underfilling due to the severe spattering caused by extremely high power density. Such a bright laser with high power density can be effective in the production of deep weld beads.

The above example with the slab-type bright laser may show exceptional penetration depths. In general, in welding with LD-pumped CW YAG lasers, the welding phenomena and imperfection formation tendencies are the same as those with lamp-pumped CW YAG lasers. The penetration depends mostly upon the power and its density.[35]

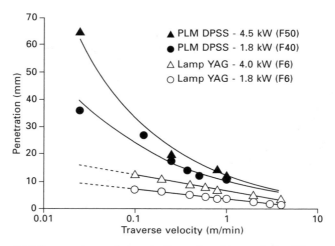

6.11 Comparison of penetration of welds made by LD-pumped slab and lamp-pumped rod YAG lasers at various speeds. DPSS = diode pumped solid state.

6.2.6 Disk lasers

Yb:YAG disk lasers are relatively new and expected to have higher power, higher efficiency and higher intensity (brightness).[35] Disk laser systems of 1 and 4 kW are commercially available and are delivered through 150 and 200 μm diameter fibers, respectively.[6,7,35] The principle of the laser system are shown in Fig. 6.12. Figure 6.13 gives a comparison of weld penetration between 4 kW rod and thin disk lasers.[7] It is confirmed that deeper penetration is obtained with a disk laser of higher power density at higher speeds. Even a 1 kW class laser produces a deep keyhole-type weld bead with extremely

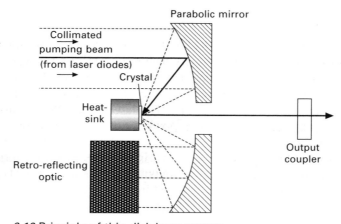

6.12 Principle of thin disk laser system.

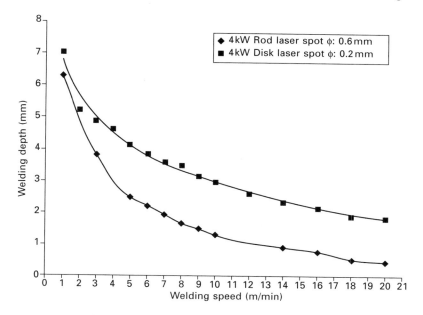

6.13 Comparison of weld penetration between rod and thin disk lasers at 4 kW.

narrow width in stainless steel and aluminum alloy. It is expected that such a laser with superior beam quality will be utilized in place of high power CO_2 or lamp-pumped YAG lasers. Moreover, a thin disk laser could be used as a heat source for remote/scanner welding with a robot.

6.2.7 Fiber lasers

Fiber lasers have good beam quality and are now recognized as being highly efficient, bright and high-power lasers. High power lasers for welding are being rapidly developed using pumping system and fiber coupling, as shown in Fig. 6.14.[38] Deep weld beads can be produced with the fiber laser as well as with the LD-pumped rod YAG laser. The laser at 6.9 kW can provide a deeply penetrated weld at high speed.[39] Moreover, it is possible to use fiber lasers as heat sources for remote or scanner welding in conjunction with robots, in place of high quality (slab) CO_2 lasers.[17] Fiber laser appliances of 10 kW or more are available and those of 100 kW power levels are scheduled.

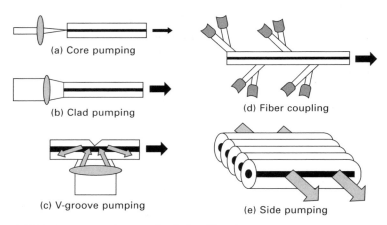

6.14 Various pumping methods for fiber laser systems.

6.3 New areas of research in laser welding

6.3.1 Laser welding of steels, Zn-coated steels and stainless steels

Low carbon steel sheets of approximately 3 mm or less in thickness are subjected to tailored blank welding with CW lasers. In order to achieve high formability and stamping ability of a steel weld, or in the use of low-carbon sheets and wires, a slight increase in heat input by lower-speed laser welding or hybrid welding is adopted. A beam shaping, scanning or spinning, and/or high speed laser welding could be used. These methods can suppress or narrow a hard weld as well as minimizing or eliminating an underfilled level of the weld bead surface.[40]

In the welding of steels with a high content of carbon, attention is paid to the occurrence of solidification cracking, and sometimes, cold (hydrogen embrittlement) cracking and the formation of hard, brittle, fragile martensite and/or cementite. It is noted in Japan that even mild steels with extremely fine grains (of about 1 μm in diameter) are hard and strong due to the microstructure-hardening effect, according to the Hall-Petch equation ($\sigma = \sigma_o + k/d^{1/2}$; where σ = yield stress; d = grain diameter; and σ_o and k = constants). They are equivalent to the properties of high tensile strength (HT) steels.[41,42] However, the problem is that the heat-affected zone (HAZ) of weld beads becomes soft due to the coarsening of grains and disappearance of strain hardening. Consequently, rapid cooling during laser welding will maintain the mechanical properties of the HAZ by the suppression of grain coarsening and the formation of hard martensite.[41,42] Laser welding of HT steels is also under investigation for such items as cars and pipes worldwide.[43,44] In HT steels, the low temperature toughness of weld joints can be maintained by decreasing the hardnesses of weld beads and the HAZ. It should be noted that it is rather difficult to evaluate the toughness of a welded joint with an extremely narrow hardening zone.

Zn-coated steels are used in industry because of low prices and higher corrosion resistance. Sound laser weld beads with good surface appearance are easily produced in lap welding with a correct gap (about 0.1 mm depending upon the Zn layer thickness[45]) as well as in butt-joint welding.[46] In the case of a lap joint with a wide gap (for example, 0.5 mm for 1 mm thick sheets[47]), weld beads are formed separately in the upper and lower sheets. Thus the control of a gap or its absence are desirable for the production of a sound weld. It is, however, known that evaporated Zn causes spatters or porosity easily in laser lap welding of steel sheets with a rather thick Zn-coated layer without a gap.[45] The formation and characteristics of weld beads were investigated using ultra-high speed video and microfocused X-ray transmission imaging system and monitoring signals, as shown in Fig. 6.15.[46,47] In welding of lapped joints without a gap between sheets at low speeds, some bubbles of Zn vapors come into the molten pools from the peripheral lapped part of the HAZ, resulting in large pores or wormholes. On the other hand, at high welding speeds in sheets without a gap, spattering occurs easily, resulting in heavily underfilled weld beads. It has been reported that sound weld beads can be produced in laser lap welding without a gap under the following conditions: with an elongated beam,[43,48] by properly tilted beam irradiation at a high power,[49] under the irradiation conditions of optimum pulse width and repetition,[46] by using Cu insert sheet[50] equivalent to the use of Cu-Si wire in laser blazing,[11,12] or by the selection of the hybrid welding with a laser arc.[51,52] It is important to reduce the harmful effect of Zn vapor affecting the melt in the molten pool.

Steels are generally welded in an inert gas shielding. It has been shown that in the laser welding of normal carbon and HT steels with high-affinity alloying elements with oxygen CO_2 gas shielding is more effective in reducing porosity than is Ar or He gas shielding.[53] Also, austenitic stainless steels can be welded where the shielding gases are Ar, He, N_2, or a mixture of these.[19,20,54] Pores are easily formed in a deeply penetrated weld bead made with a CO_2 or YAG laser with Ar or He shielding but are almost eliminated in N_2.[20,54] *In-situ* observation of X-ray transmission reveals that the generation of bubbles can be suppressed and some bubbles can shrink and disappear in the molten pool in N_2 shielding gas.[20] The reason for reduced porosity is attributed to the higher solubility of N in the molten pool and the easier combination of N with Cr vapor during welding. The N content in the weld fusion zone is higher with a CO_2 laser than with a YAG laser due to the formation of N plasma.[54] Solidification cracks are absent in YAG laser welds but are present along grain boundaries of the austenite phase under some conditions with a CO_2 laser in N_2 gas. Attention should be paid to the use of N_2 gas in CO_2

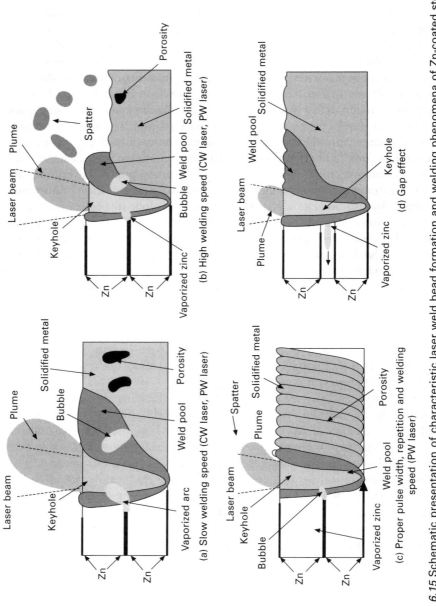

6.15 Schematic presentation of characteristic laser weld bead formation and welding phenomena of Zn-coated steel sheets.

laser welding of austenitic stainless steels. This is because the absorbed N is such a strong austenitizing element that the solidification process varies from the primary solidification of the ferrite phase to that of austenite phase; consequently, microsegregation of P and S along grain boundaries increases to lower the solidification temperature and to widen the solidification brittleness temperature range (BTR) leading to enhanced cracking sensitivity.[54]

In pulsed YAG laser spot welding of stainless steels, the microstructure at room temperature, i.e. the ferrite-to-austenite phase ratio of the weld metals, is quite different from that of normal TIG weld fusion zones, as shown in Schaeffler diagram in Fig. 6.16.[55,56] For example, although TIG weld fusion zones show duplex microstructure with about 5 % and 30 % residual delta(δ)-ferrite content, the weld metals produced with a pulsed YAG laser with several ms irradiation duration generally exhibit almost fully austenitic and ferritic microstructure, respectively. This is interpreted in terms of rapid solidification and the subsequent rapid cooling.[55,56] In the case of austenitic stainless steels producing TIG weld metals in AF mode (the primary austenite and subsequent eutectic or peritectic ferrite solidification process), in pulsed or high-speed CW laser weld metals, a fully austenitic microstructure is formed. This is caused either by the primary solidification of the austenite phase without subsequent transformation,[55] or by primary ferrite solidification with a reduced level of microsegregation due to the rapid solidification

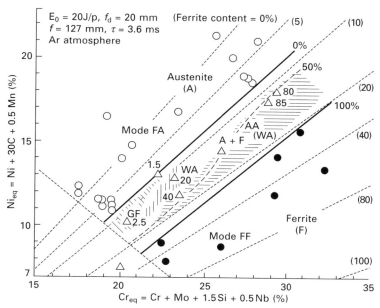

6.16 Ferrite contents and microstructure of pulsed laser spot welds shown in Schaeffler diagram.

174   New developments in advanced welding

effect and the complete solid-state transformation from the ferrite to the austenite during subsequent rapid cooling.[56] There is a possibility of solidification cracking only in the case of the primary solidification of the austenite phase.

6.3.2 Laser welding of aluminum or magnesium alloys

Aluminum and magnesium alloys have received much attention due to their light weight, attractive surface appearance, and other suitable properties. They are widely used in many industries – such as electrical, electronics and transportation.

Laser welding of aluminum alloys is generally difficult because of high light reflectivity (low coupling efficiency), high thermal diffusivity (conductivity), easier formation of welding defects such as porosity in deeply penetrated welds and hot cracking in pulsed spot welds.[21,22,24] Melting is enhanced by utilizing a high power density laser or by forming a AlN phase in N_2 shielding gas during CO_2 laser welding.[57,58] Moreover, deeper penetration can be obtained in those alloys with a larger content of volatile elements, such as Mg, Zn and Li, as shown in Fig. 6.17.[59] However, porosity can easily occur in aluminum alloy weld metals with a higher content of magnesium. In the case of high power laser welding of wrought aluminum alloys, many bubbles are generated from the bottom tip of a keyhole, resulting in the

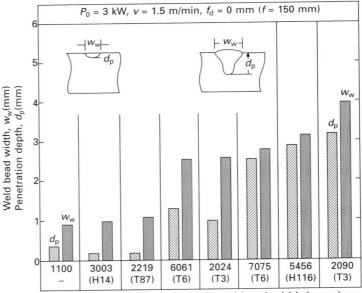

6.17 Comparison of penetration depth and bead width in various aluminum alloys.

New developments in laser welding 175

Table 6.2 Q-mass analyses of porosity inside gases, showing gas compositions (mass %) in pores formed in laser welding of A5083 alloy

Laser kind	Shielding gas	Power	Ar	He	H_2	Others
CO_2 laser	Ar	5 kW	41	–	59	–
CO_2 laser	He	10 kW	0.6	86.8	12.6	–
CO_2 laser	He	10 kW	0.6	95.9	3.3	0.2 N_2
YAG laser	He	3 kW	–	99.2	0.6	0.2 N_2

formation of porosity. In aluminum alloys more bubbles are generated from a keyhole tip and float upwards depending on the melt flows in the molten pool. Gas constituents inside the large pores in A5083 alloy welded with a CO_2 and with a YAG laser under given conditions were analyzed with a drilling Q-mass system in a vacuum and the results are summarized in Table 6.2.[60] The shielding gas is mainly present in the pores and both hydrogen gas and nitrogen are detected in the pore atmosphere. The hydrogen content is high in the CO_2 laser welds and increases with time before analysis takes place; hydrogen must invade the pores by diffusion during and after welding. It is therefore concluded that bubbles leading to porosity are formed by intense evaporation at the keyhole front near the bottom of the molten pool and the shielding gas is entrained into the keyhole and bubbles result.

In an inert shielding gas with a small amount of H_2, a great number of small pores are formed in aluminum alloy weld beads made with a CO_2 laser. Therefore, the use of a pure inert gas and the polishing of the plate surface, where the oxygen and hydrogen content are normally high, can decrease small-sized porosity caused by hydrogen. Several procedures such as correct pulse modulation, using moderate power density, twin laser beams, hybrid welding process and full penetration welding can reduce porosity in weld fusion zones of wrought aluminum alloys.[60] The formation mechanisms and porosity prevention are the same for magnesium wrought alloys. In some cases porosity occurs so easily that it is difficult to reduce or eliminate it. Such cases include the laser welding of casting, die-casting, thixomolding and working with powder metallurgy products of aluminum and magnesium alloys. These often show small-sized porosity, blowholes, hydrogen or oxygen gas-enriched areas and other such features.

The susceptibility to hot cracking such as solidification cracking in the weld metal and liquation cracking in the HAZ can be expressed as a function of alloy content, as indicated in Fig. 6.18.[60] Laser welding processes have a marked effect on a tendency towards hot cracking. Solidification cracking occurs easily in pulsed YAG laser spot weld metals and in CW laser high-speed weld beads of aluminum and magnesium alloys, while weld beads

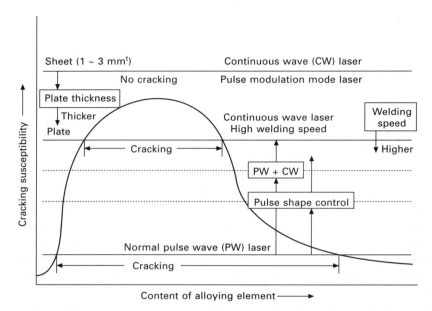

6.18 Correlation of various laser welding processes to hot cracking susceptibility as a function of alloying element content.

without cracks can be produced in welding of 2 or 3 mm thick sheets with a CW laser. These are interpreted through the formation of a wider solid–liquid mushy zone and the rapid loading rates of augmented tensile strains to grain boundaries.

The mechanical properties, such as tensile and fatigue strength, of aluminum alloy joints are chiefly affected and degraded by the size of large pores, cracks, and the degree of underfilling. Furthermore, softening of HAZ due to annealing and overaging phenomena during welding can also decrease the mechanical properties of work-hardening and age-hardening materials.[22,61,62]

6.3.3 Laser welding of dissimilar materials

Welding or bonding of dissimilar metallic materials is receiving much attention because of the great demand for high quality and high performance industrial products. However, fusion welding of dissimilar alloys is notoriously difficult due to the ready occurrence of cracking in the intermetallic compounds formed. Nevertheless, some good results have recently been obtained in laser lap-joint or lap and butt joint one-pass welding of dissimilar materials such as aluminum alloy and steel, as shown in Fig. 6.19.[63–66] Intermetallic compounds should be very thin and their thicknesses are controlled by melting aluminum alloy sheet only at confined depths in the plate. When the bonding

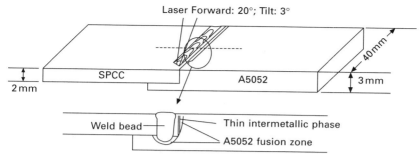

6.19 One pass welding of a butt and lap joint configuration for welding aluminum alloy and steel.

area is wide enough, the mechanical properties of the joints are so good that the fracture occurs in a base metal in the tensile test[63]. Moreover, the possibility of laser welding of SPCC or Type 304 to Mg alloy is confirmed when the laser is shot on the steel side without melting of the joint interface so that the heat of the laser weld fusion zone can melt Mg alloy plate.[67] The feasibility of laser welding of copper to steel or Ti, and of other high temperature metals to normal materials has been investigated.

6.3.4 Laser welding of plastics

Laser welding of plastics has been actively investigated with the evolution of diode lasers all over the world. The plastics are readily joined using diode, YAG and fiber lasers of low power under defocused conditions, as shown in Fig. 6.20, since the absorption and transparency are easily controlled by the concentration of such substances as carbon black.[68,69] The melting points of plastics are generally lower than those of metals and the processing temperature ranges are narrow because of low evaporation temperatures. LD welding of plastic parts is applied practically in mass production by the Toyota Motor Corp. and its group.[69] LD welding of the intake manifold, canister, cutoff valves in the fuel tank and fog lamps are examples.[69] In addition, almost transparent plastics can be lap-bonded by using slightly higher temperatures generated at the joint-bonded area and using jig plates with a larger radiation heat in CO_2 laser welding.

6.3.5 Laser welding phenomena and porosity formation

The understanding of welding phenomena and porosity formation mechanisms is important in laser welding. This means that plume and keyhole behavior, melt flows and bubble formation and porosity formation are relevant. Their observations have been taken during spot and bead welding with pulsed

178 New developments in advanced welding

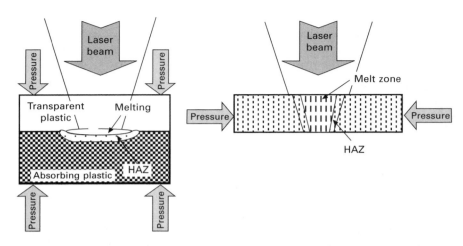

6.20 Laser welding processes for plastics (left: lap joint; right: butt joint).

YAG laser and CW YAG and CO_2 lasers under various conditions using high speed videos and X-ray transmission systems.[21,22,24,70]

In spot welding, the rapid collapse of a deep keyhole and subsequent rapid solidification are causes of bubble and porosity formation, as shown in Fig. 6.21.[71] The porosity can be suppressed by the application of pulse shaping and the addition of correct tailing power.[24-26,71]

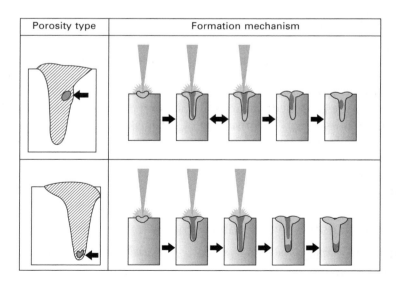

6.21 Porosity in laser spot welding and porosity formation mechanism.

For bead welding, welding phenomena and porosity formation are schematically summarized in Fig. 6.22.[70] Keyhole behavior, melt flows and bubble and porosity formation depend apparently upon the kind of material as well as on welding conditions such as laser power and welding speed. Such keyhole behavior and melt flows are best understood by considering the factors determining evaporation. Such factors are the different locations of laser–material interaction and keyhole collapse, the content of volatile elements and such physical properties as the vaporization temperature and the surface tension. At low welding speeds, a deep keyhole is liable to collapse, a laser beam is shot on the liquid wall of the collapsed keyhole, and consequently the downward melt flow along the keyhole wall is induced by the recoil pressure of evaporation. Moreover, intense evaporation takes place at the front wall of the keyhole and thereby many bubbles are generated from the keyhole tip. As the welding speed increases, vapors generated from the keyhole wall are more strongly ejected upwards through the keyhole inlet. As a result, upward melt flow is induced near the keyhole inlet and downward flow near the keyhole tip is reduced. A bubble moves a short distance and the formation location of porosity is limited to the area near the bottom. As the welding speed increases considerably, the keyhole becomes narrower and shallower, bubbles become smaller and consequently the number of pores

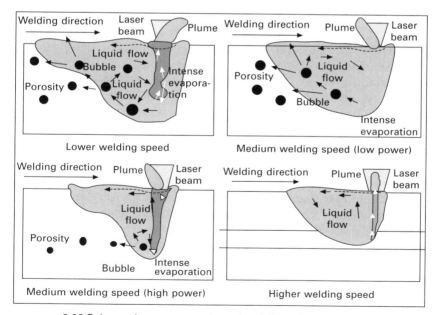

6.22 Schematic representation of welding phenomena: plume, keyhole, melt flows and bubble and porosity formation under various conditions.

decreases and the bubbles do not reach the surface of the molten pool. This means that the formation ratio is high and volume of porosity large at a certain welding speed. The formation of bubbles and of porosity are almost similar in YAG and CO_2 laser welding.[19–22,72,73]

Certain laser welding conditions reduce porosity or large pores. Such conditions include proper pulse modulation, moderate power density, a forwardly declined (tilted) laser beam, a very low speed or a high speed, the weaving method, a twin-spots laser, full penetration welding, vacuum welding, the use of a tornado nozzle and hybrid welding with a TIG or MIG arc at high currents.[19–22,70–73]

6.4 Advances in laser welding processes

6.4.1 Remote laser welding

Remote or scanner laser welding is a highly efficient bonding process and CO_2 lasers up to 6 kW power are available with focal lengths up to 1.5 m.[16,17] A system using a CO_2 laser of extremely high beam quality has been developed to weld car parts, where a beam with long focal length is deflected by scanner optics, positioned and moved over the workpiece at high speeds. This welding can be completed in a much shorter time than can processes involving general resistance spot welding and laser seam welding.[16] Attention should be paid to such items as complex clamping devices and accessibility, the effect of laser beam inclination angle to the workpiece on penetration and the gas shielding situation.[17] It is anticipated that new concepts including robot guided scanner welding or flexible remote welding will allow flexible manufacturing for a wide range of applications, as shown in Fig. 6.23.[18] Fiber and disk lasers of high beam quality are heat source candidates for robot-coupled systems.

6.4.2 On-line or in-process monitoring during laser welding and adaptive control

On-line or in-process monitoring and feedback or adaptive control are necessary and have been intensively investigated to produce high quality welds.[74–86] In the monitoring process a reflected laser beam, light emission from the plasma/plume and/or molten pool, etc., sound from the plasma or keyhole inlet, ultrasonic or acoustic sound from the metal inside, the plasma potential between the plate and the nozzle or laser-induced plume and other phenomena are investigated as signals in conjunction with penetration or welding defects. Monitoring/adaptive control systems are available; they utilize imaging observation of a keyhole and a molten pool and the reflection light of another laser beam such as LD and He–Ne laser from the butt-joint edge to be

New developments in laser welding   181

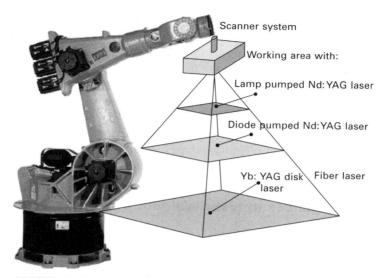

6.23 Effect of beam quality on working area with a robot.

welded or from the weld bead surface profile after welding. These are shown in Figs. 6.24 and 6.25.[74] Coaxial imaging observation can judge a gap and also full or partial penetration in sheets and plates. Adaptive or feedback control systems are applied by varying either laser power or the welding speed on the basis of the data after detecting a butt-joint or a lap-joint gap during continuous or stitch welding with a high power laser. According to recent research involving aluminum alloys, it is interpreted that large spatters increase the heat radiation signal, and the melt-removed surface after spattering increases the beam reflection in the upward and subsequent inclined direction.[81] The correlation between the reflected beam intensity and the heat radiation from the molten pool can be used to interpret full and partial penetration in the sheet.[81]

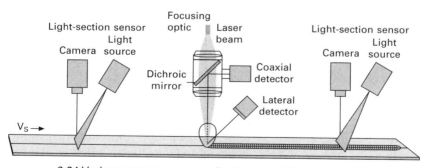

6.24 Various sensor systems for laser welding.

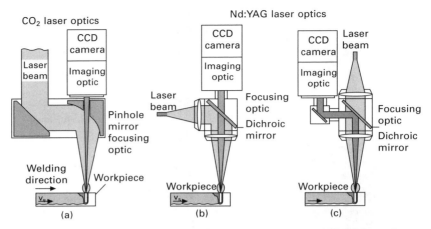

6.25 Schematic drawing of different sensor set-ups for CO_2 and Nd:YAG laser systems.

Underfilling, which degrades strength and formability, is also detected by the laser line interference method or plume light intensity.[11, 83] Some on-line and off-line systems are used in the production lines of tailored blanks or aircraft panels, as shown in Fig. 6.24 and 6.26,[77] respectively.

Moreover, in YAG laser spot welding of thin sheets, the lap-welded areas sometimes vary with samples and a through-hole defect is easily formed in the upper thin sheet during laser irradiation in some samples. Examples of the formation of a normal weld and non-bonded weld with a through-hole are illustrated schematically in Fig. 6.27.[85] To produce sound laser partial-penetration lap-joint welds consistently, a new procedure of in-process monitoring and adaptive control has been developed for laser micro-spot lap

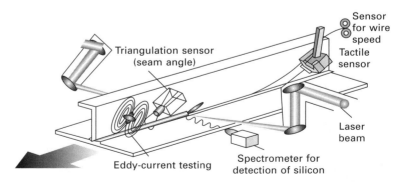

6.26 On-line technique for inspection of laser welded panels of aircraft.

New developments in laser welding

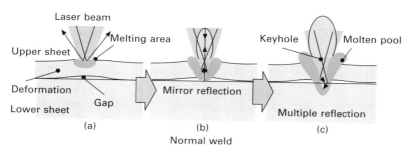
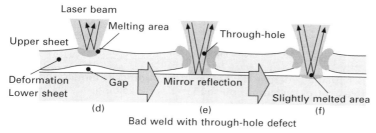

6.27 Schematic representation of formation mechanisms of normal weld (a, b, c) and bad weld with a through-hole defect (d, e, f).

welding of A3003 aluminum alloy sheets of 0.1 mm and 1 mm in thickness. The system is shown in Fig. 6.28.[85] The reflected laser beam and the radiated heat from the welding area are revealed to be effective as in-process monitoring signals in detecting melting, keyhole generation and through-hole formation in the upper sheet during laser irradiation. Laser pulse duration and peak power can be controlled at every 0.15 ms interval during the laser spot welding on the basis of the heat radiation signal. Sound partially penetrated spot welds are produced in all samples subjected to laser lap welding under the two proposed in-process monitoring and adaptive control methods. An in-process repairing technique has also been developed, during which the laser power is increased so as to melt further the lower sheet for bonding the two sheets after the detection of through-hole defect formation during spot welding, as shown in Figs 6.29 and 6.30.[85] Such in-process monitoring and adaptive control systems have been developed to produce consistently sound partially and fully penetrated lap spot welds.[85,86]

6.4.3 Laser-arc hybrid welding

Hybrid welding with CO_2, YAG or LD lasers and TIG, MIG, MAG (metal active gas) or another heat source has been receiving considerable attention regarding such factors as depth of penetration, higher welding speeds, wider gap tolerance and lower porosity.[87–100] The production of a compact head is

184   New developments in advanced welding

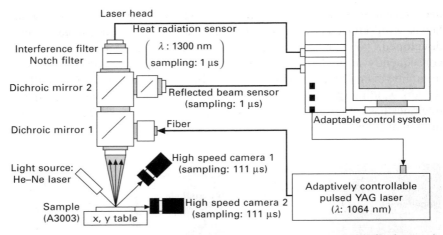

6.28 Schematic experimental set-up of an in-process monitoring and adaptive control system for pulsed laser spot welding.

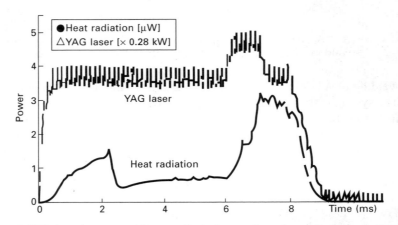

6.29 Pulse shape used for on-site hole repair and monitoring results of heat radiation during laser spot welding of A3003 alloy under adaptive control.

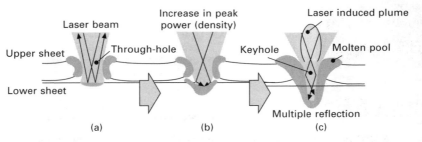

6.30 Schematic mechanism for in-process repairing of a through-hole defect during spot welding.

necessary in industrial applications, as shown in Fig. 6.31,[89–92] and coaxial heads with TIG electrode/MIG wire and a YAG laser beam have been developed.[90] Laser welding is carried out with a focused beam, and consequently an underfilled laser weld bead is easily formed in any gap joint however slight; use of a filler wire for reduction in underfilling renders welding speed slow. The effect of laser–MIG hybrid welding on gap tolerance has been frequently demonstrated and is shown in Fig. 6.32.[91] The mixing of filler wire components also takes place more completely in hybrid

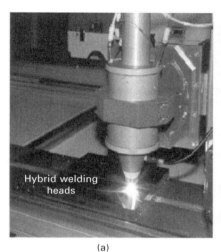

(a)   (b)

6.31 Coaxial TIG–YAG (a) and MIG–YAG (b) hybrid welding heads.

Material: S 690 QL
Speed: 3.0 m/min
Laser power: 3 kW
Welding current: 135 A

| Nd:YAG–laser welding gap 0.2 mm | TIG-Nd:YAG–laser welding gap 0.2 mm | MIG-Nd:YAG–laser welding gap 0.2 mm |
|---|---|---|//

6.32 Comparison of cross-sections of weld beads produced by laser, TIG–YAG and MIG–YAG hybrid welding.

welding than it does in laser welding due to the effect of arc electromagnetic convection.

The effects of various welding parameters on weld penetration and porosity formation in hybrid welding with CO_2, YAG or diode laser and TIG, MIG and MAG have been reported.[96–100] It is understood that weld bead penetration and geometry are greatly affected by the electrode-to-laser beam target distance and the welding direction. In YAG–TIG hybrid welding, the deepest penetration can be obtained at short distances of 1 to 2 mm; however, the penetration becomes equal to or shallower than that of laser welds at distances of 5 to 9 mm. On the other hand, in TIG–YAG hybrid welding, the penetration depth is always deeper than that of laser welds and the deepest penetration is attained at the distance of about 5 mm. The effect of oxygen in air on the surface tension and arc constriction in the normal Ar flow is revealed to be the same as that of sulfur in arc and hybrid welding of stainless steel. The formation of the deepest weld bead is attributed to several superimposed downward flows along the keyhole wall. These are caused by recoil pressure against the keyhole wall or collapsing keyhole, marangoni (surface tension driven) convection from low to high temperatures, the arc constriction and electromagnetic convection due to a high content of O_2, in addition to the keyhole depth. On the other hand, shallow penetration is due to the collision between the downward flow along the keyhole wall caused by the recoil pressure and the other downward flow near the central molten pool caused by marangoni convection and the electromagnetic convection due to TIG arc constriction.[99]

TIG–YAG hybrid welding phenomena in air at 100 and 200 A are schematically shown in Fig. 6.33.[99] The diameter at the upper part of the keyhole becomes larger with increasing arc current. At 100 A, a keyhole was slightly larger and deeper than that produced in laser welding; then the downward flow of the melt near the keyhole wall became dominant as the

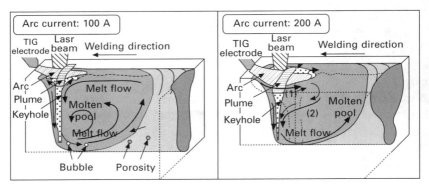

6.33 Schematic illustration of TIG–YAG hybrid welding phenomena in air at 100 A and at 200 A.

Welding speed v. 40 mm/s (thickness: 4 mm)

Arc current (A)	0 (YAG single welding)	60	120	180	240
Cross-section			Porosity		3 mm

Material: A5052 (4 mm), YAG laser power, P_1: 3.1 kW, Shielding gas: Ar 30 l/min, Defocused distance: 0 mm, Torch angle, α: 30°, Welding direction: YAG–MIG, Distance, d: 2 mm

6.34 Effect of MIG arc current on penetration of YAG–MIG hybrid welding at 40 mm/s.

melt flowed from the keyhole tip to the rear along the molten pool bottom. The latter flow deepened the bottom of the molten pool, leading to the deeper weld bead. Under these conditions large bubbles are often generated to form larger-sized pores. On the other hand, at 200 A, the molten pool surface was depressed, the keyhole diameter near the top of the surface was larger, and other fast melt flows were observed around the keyhole near the surface, resulting in the formation of wider bead widths in the upper part. Moreover, no or reduced generation of bubbles was observed, leading to no or reduced porosity at high arc currents.[99]

In the welding of aluminum alloys, hybrid welding with YAG–MIG heat sources can produce superior weld beads without undercuts giving a good appearance to top and bottom surfaces in comparison with those produced by laser welding. MIG–YAG welds generally appear to be slightly deeper and larger than are YAG–MIG welds. The surfaces of YAG–MIG welds always appear better than welds MIG–YAG.[100] Figure 6.34 shows cross-sectional YAG–MIG weld beads made at 40 mm/s as a function of MIG arc current.[100] The weld beads become larger and deeper with an increase in the MIG current.

Porosity is reduced in A5052 hybrid weld beads at the laser power of 3 kW and at a high MIG current of 240 A.[100] Welding phenomena, molten pool geometry, melt flows inside the pool, bubble and porosity formation are illustrated schematically in Fig. 6.35.[100] At 120 A, many bubbles are generated and trapped by the solidifying front, resulting in porosity formation. On the other hand, at 240 A, some bubbles are generated and all bubbles disappear from the concave surface of the molten pool, resulting in no porosity.

It is interesting to know that the concave molten pool surfaces induced at high TIG and MIG currents suppress bubble formation due to the formation of a more stable keyhole in stainless steels or act as a disappearance site for bubbles in aluminum alloys.

188 New developments in advanced welding

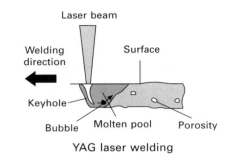

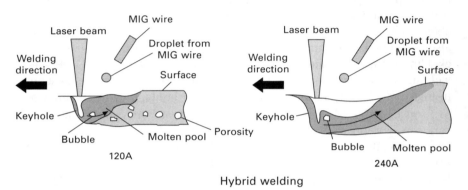

6.35 Schematic representation of welding phenomena, molten pool geometry, melt flows inside the pool, bubble and porosity formation during laser and YAG–MIG hybrid welding.

6.5 Applications of laser welding

6.5.1 Automobile industry

Car manufacturing companies use large numbers of 2.5 to 6 kW class lasers to weld various parts, tailored blanks and bodies-in-white.[6,12,101] Two laser machines can be separately or simultaneously used, and this system is effectively operated where one laser robot can cover the whole area while the other is out of order. Steels, zinc-coated steels or aluminum alloys are welded with several kW class CO_2 lasers, or 2.5 to 4 kW lamp-pumped or LD-pumped YAG lasers.[6,12,101] Laser blazing as well as laser lap-welding is considered a promising method in welding Zn-coated steels.[101] Most car companies are also interested in remote laser welding systems.[16–18]

6.5.2 Steel industry

The steel industry introduced high power lasers to weld thin and thick plates in the 1980s and 1990s. Steel and stainless steel sheets or plates are welded to produce coils, long plates or pipes. Two sets of 45 kW CO_2 lasers are operated to weld 30 to 40 mm thick hot slabs at about 1300 K, where laser beams delivered by mirrors are used to weld moving plates.[10] This advanced system has been developed in conjunction with the development of many other peripheral technologies.

6.5.3 Heavy industry

Laser welding of high quality and high productivity is used in nuclear and thermal power plants of heavy industry. High power lasers are utilized to weld thin stainless steel sheets of 1 mm thickness as well as thick plates of about 16 mm.[29,30] Laser welding of Co-base or Ni-base alloys for turbines is also performed. Laser underwater welding and/or repairing as well as laser cladding and laser peening are also investigated in heavy industry. Underwater welding can be carried out with a fiber-delivered YAG laser system, as shown in Fig. 6.36.[102] High power diode lasers, YAG lasers, disk lasers and fiber

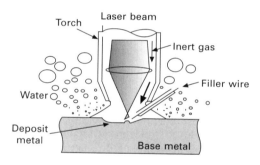

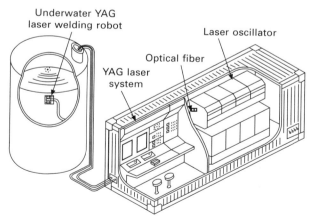

6.36 Schematic representation of underwater welding and a fiber-delivered YAG laser system.

lasers are possible heat sources for welding, repairing and cladding. Recently laser-arc hybrid welding has also received attention for welding of thick plates in ships.

6.5.4 Electronic and electrical industries

The electronic and electrical industries make considerable use of pulsed YAG lasers as micro-welding sources. Battery cell cases of A3003 aluminum alloy are welded with a pulsed YAG laser and this alloy is essentially resistant to hot cracking. However, crater solidification cracks are present in the last spot weld, and this is a very important problem in the behaviour of aluminum alloys.[103] There is an increasing demand for the welding of small parts as well as a variety of materials, such as stainless steels, Al, Cu, Ti, precious metals or alloys of Au, Ag and Pt, high temperature alloys of W, Mo, Ta and Nb, and dissimilar materials. Therefore the use is anticipated of pulsed normal or second harmonic generation (SHG) YAG lasers, disk lasers and fiber lasers of high beam quality and single mode fiber lasers and SHG YAG lasers are especially expected to be developed further. A recent example is an IC chip welded with an SHG YAG laser.[9] Interesting and promising applications are the micro-spot welding of copper sheets.

6.6 Future trends

LD-pumped solid-state lasers, such as disk and fiber lasers, are being investigated regarding higher power, higher beam quality, higher efficiency and fiber delivery. They are also expected to act as heat sources in place of CO_2 lasers for remote/scanner laser welding because they are more amenable to robot and fiber delivery. Such lasers also have advantages in easier operation, higher productivity and cost reduction. On the other hand, there are merits in welding with laser-arc hybrid heat sources and SHG pulsed YAG lasers, or short wavelength lasers to sustain interest in them. There is no one laser suitable for all tasks available at present. So any laser should be selected on the basis of a good understanding of its performance and applications.

Complicated laser welding phenomena and the mechanisms of the formation of imperfections in laser welding have been clarified in conjunction with the development of observing and measuring instruments. Interpretation of results will progress quickly by utilizing simulation techniques for welding phenomena using data on the physical constants of materials which are becoming increasingly accurate. Remaining problems will be resolved gradually.

For these reasons, laser welding is increasingly applied to bond or join many similar and dissimilar materials. In some cases there is increasing necessity for in-process monitoring and adaptive control. Laser systems are still extremely expensive in comparison with arc heat sources, which necessitates the further development of the laser welding system in every field.

There is an increasing necessity for research projects of worldwide scale. Here many experts in the field of lasers, materials, instruments, products, and applications, gathering as representatives for several companies, will collaborate. Experiments and research projects performed by only a few workers in one group or company are inefficient; the development of intelligent laser systems with high performance demands much time and money. Advances in lasers and laser welding processes depend upon personnel and project budgets.

Laser welding technology is being intensively investigated together with the development of new lasers of higher beam quality under small or large projects worldwide, especially in Germany. Much fruitful research and its industrial applications are published or announced in journals and international conferences each year. The trend of development in lasers and welding processes with lasers or hybrid heat sources will continue in each industrial field as long as laser and hybrid welding are recognized as high technology. In future, every research and development activity of laser welding should be performed from the viewpoint of environmental protection and safeguards.

6.7   References

1. Maiman T.H., 'Stimulated optical radiation in ruby masers', *Nature*, 1960 **187** 493.
2. Katayama S., 'Current states and trends of laser welding', *Journal of High Temperature Society*, 2001 **27**(5) 190–202 (in Japanese)
3. O'Neil W., Sparkes M., Vamham M., Horley R., Birch M., Woods S. and Harker A., 'High power high brightness industrial fiber laser technology', *Proc. of the 23rd Int. Congress on Applications of Lasers & Electro-Optics (ICALEO) 2004*, San Francisco, LIA, 2004, Fiber & Disc Lasers Session, 1–7 (CD)
4. Bachmann F., 'The impact of laser diodes to solid state lasers and materials processing applications', *Proc. of the 61st Laser Materials Processing Conf.*, Osaka, JLPS, 2004 **61** 16–29
5. Lossen P., 'Data from Fraunhofer Institut fur Lasertechnik (Private Communication)' used by: Tsukamoto S., 'Laser welding', *Journal of the Japan Welding Society*, 2003 **72**(1) 16–21 (in Japanese)
6. Mann K., 'The disk laser, an advanced solid state laser technology for new application', *Proc. of the 61st Laser Materials Processing Conf.*, Osaka, JLPS, 2004 **61** 1–10
7. Morris T. and Mann K., 'Disk laser enables application advancements', *Proc. of the 23rd ICALEO 2004*, San Francisco, LIA, 2004 Fiber & Disc Lasers Session, 24–7 (CD)
8. Shiner B., 'Fiber lasers and their application', *Proc. of the 61st Laser Materials Processing Conf.*, Osaka, JLPS, 2004 **61** 11–15
9. Uchida T, 'Spot welding of copper with SHG laser' (Private Communication, Miyachitechnos)
10. Alder H., 'Recent development and applications in body-in-white laser joining of

European car manufacturers', *Proc. of 59th Laser Materials Processing Conference*, Nagoya, JLPS, 2003, **59** 36–44
11. Shirai M., 'Laser welding on trough panel (3 dimensional body part)', *Proc. of the 55th Laser Materials Processing Conf.*, Nagoya, JLPS, 2002 **55** 1–6
12. Mori K., 'Application of laser welding for body in white', *J. of the Japan Welding Society*, 2003 **72**(1) 40–3 (in Japanese)
13. Minamida K., 'High power laser applications in Nippon Steel Corporation', *Proc. of SPIE First Int. Sym. on High Power Laser Macroprocessing*, Osaka, JLPS, 2002 **4831** 402–10
14. Roland F., Laser welding in shipbuilding – chances and obstacles', *Proc. 8th Sym. on Laser Materials Processing*, KIMM, Korea, 1997 237–52
15. Ito M., Takada M., Yanagishima F., Kawai Y., Yokozawa F., Nakahara H. and Sasaka H., '10kW laser beam welder for stainless steel processing line', *Proc. LAMP '87*, Osaka, HTSJ & JLPS, 1987 535–40
16. Lingner M., 'The remote welding in industrial applications', *Proc. of 59th Laser Materials Processing Conference*, Nagoya, JLPS, 2003, **59** 21–35
17. Vollertsen F., Seefeld T. and Grupp M., 'Remote welding at high laser power', *Proc. of the 22nd ICALEO 2003*, Jacksonville, LIA, 2003, 376–85 (CD)
18. Klotzbach A., Morgenthal L., Schwarz T., Fleischer V. and Beyer E., 'Laser welding on the fly with coupled axes systems', *Proc. of the 20th ICALEO 2001*, Jacksonville, LIA, 2001 (CD)
19. Katayama S., Seto N., Kim J.D. and Matsunawa A., 'Formation mechanism and reduction method of porosity in laser welding of stainless steel', *Proc. of ICALEO '97*, San Diego, 1997 **83**(2) Section G 83–92
20. Seto N., Katayama S. and Matsunawa A., 'High-speed simultaneous observation of plasma and keyhole behavior during high power CO_2 laser welding: effect of shielding gas on porosity formation', *J. of Laser Applications*, 2000 **12**(6) 245–50
21. Katayama S. and Matsunawa A., 'Formation mechanism and prevention of defects in laser welding of aluminium alloys', *Proc. of CISFFEL*, Toulon, SI, 1998 **1** 215–22.
22. Katayama S. and Matsunawa A., 'Laser weldability of aluminum alloys', *Proc. of 43rd Laser Materials Processing Conference*, Osaka, JLPS, 1998 **43** 33–52 (in Japanese)
23. Washio K., 'States of the arts and trends of solid-state lasers for materials processing', *Journal of High Temperature Society*, 2001 **27**(5) 176–85 (in Japanese)
24. Matsunawa A., Katayama S., Mizutani M., Ikeda H. and Nishizawa K., 'Fusion and solidification characteristics in pulse-shaped YAG laser welding', *Proc. of 5th CISFFEL*, La Baule, SI, 1993 **1** 219–26
25. Katayama S., Kohsaka S., Mizutani M., Nishizawa K. and Matsunawa A., 'Pulse shape optimization for defect prevention in pulsed laser welding of stainless steels', *Proc. of ICALEO '93*, Orlando, LIA, 1993 **77** 487–97
26. Mizutani M., Tanaka K., Katayama S. and Matsunwa A., 'YAG laser spot welding under microgravity and vacuum and pulse-shaping for prevention of porosity', *Proc. of the 7th Int. Welding Symposium*, Kobe, JWS, 2001, 573–8
27. Gref W., Russ A., Leimser M., Dausinger F. and Hugel H., 'Double focus technique – influence of focal distance and intensity distribution on the welding process', *Proc. of SPIE (First Int. Sym. on High Power Laser Macroprocessing)*, Osaka, JLPS, 2002 **4831** 289–94

28. Haboudou A., Peyre P. and Vannes A.B., 'Study of keyhole and melt pool oscillations in dual beam welding of aluminium alloys: effect on porosity formation', *Proc. of SPIE (First Int. Sym. on High Power Laser Macroprocessing)*, Osaka, JLPS, 2002 **4831** 295–300
29. Tsubota S., Ishide T., Nayama M., Shomokusu Y. and Fukusaka S., 'Development of 10 kW class YAG laser welding technology', *Proc. ICALEO 2000*, LIA, Dearborn, 2000 **89** Section C, 219–29
30. Coste F., Janin F., Hamadou M. and Fabbro R., 'Deep penetration laser welding with Nd:Yag lasers combination up to 11 kW laser power', *Proc. of SPIE (First Int. Sym. on High Power Laser Macroprocessing)*, Osaka, JLPS, 2002 **4831** 422–7
31. Hayashi Y. and Ullmann C., 'High power laser diode systems and industrial applications', *Proc. of 61st Laser Materials Processing Conference*, Osaka, JLPS, 2004 **61** 47–56 (in Japanese)
32. Zediker M.S., 'Materials processing with high power diode laser systems', *Proc. of the 7th Int. Welding Symposium*, JWS, Kobe, 2001 **1** 473–8
33. Abe N., Higashino R., Tsukamoto M., Noguchi S. and Miyake S., 'Materials processing characteristics of a 2 kW class high power density direct diode laser system', *Proc. ICALEO '99*, San Diego, LIA, 1999 **87** Section A 236–44
34. Petring D., Benter C and Poprawe R., 'Fundamentals and applications of diode laser welding', *Cong. Proc. of 20th ICALEO 2001 (Laser Materials Processing Conf.)*, Jacksonville, LIA, 2001 G1601 (CD)
35. Hugel H., 'Innovative high power lasers for welding applications', *Proc. of SPIE (First Int. Sym. on High Power Laser Macroprocessing)*, Osaka, JLPS, 2002 **4831** 83–9
36. Akiyama Y., Takada H., Sasaki M., Yuasa H. and Nishida N., 'Efficient 10 kW diode-pumped Nd:YAG laser', *Proc. of SPIE (First Int. Sym. on High Power Laser Macroprocessing)*, Osaka, JLPS, 2002 **4831** 96–100
37. Koch J, 'Deep section welding with high brightness lasers', *Cong. Proc. of 20th ICALEO 2001 (Laser Materials Processing Conf.)*, Jacksonville, LIA, 2001 G1609 (CD)
38. Ueda K, 'The prospects of high power fiber lasers', *The Review of Laser Engineering*, 2001 **29**(2) 79–83 (in Japanese)
39. Thomy C., Grupp M., Schilf M., Seefeld T. and Vollertsen F., 'Welding of aluminium and steel with high-power fiber lasers technology', *Proc. of the 23rd Int Congress on Applications of Lasers & Electro-Optics (ICALEO) 2004*, San Francisco, LIA, 2004, Fiber & Disc Lasers Session, 8–14 (CD)
40. Niimi T. and Natsumi F., 'Application of CO_2 laser welding to car body', *Proc. of 28th Laser Materials Processing Conference*, Osaka, JLPS, 1992, **28** 171–84 (in Japanese)
41. Otani T., Tsukamoto S., Arakane G and Omori A, 'HAZ properties of ultra-fine grained high strength steels welded by high power laser, *Quar. J. of JWS*, 2003 **21**(2) 267–73 (in Japanese)
42. Ito R., Hiraoka K. and Shiga C., 'Characteristics of heat-affected zone of ultra-fine grained steel in ultra-narrow gap GMA welding', *Quar. J. of JWS*, 2004 **22**(3) 458–66 (in Japanese)
43. Lu F and Forrest M.G., 'Weldability comparison of different zinc coated high strength steel sheets in laser lap joining configuration without gap', *Proc. of ICALEO 2003 (Laser Materials Processing Conference)*, Jacksonville, LIA, 2003 1409 (CD)

44. Forsman T., 'Laser welding of tailored blanks', *Proc. of ICALEO 2002 (Laser Materials Processing Conference)*, Scottsdale, LIA, 2002 Section A – Welding (CD)
45. Akhter R., Steen W.M. and Watkins K.G., 'Welding zinc-coated steel with a laser and the properties of the weldment', *J. of Laser Applications*, 1991 **3**(2) 9–20
46. Katayama S., Wu Y. and Matsunawa A., 'Laser weldability of Zn-coated steels', *Proc. of ICALEO 2001 (Laser Materials Processing Conference)*, Jacksonville, LIA, 2001 Section C – Welding P520 (CD)
47. Katayama S., Takayama M., Mizutani M., Tarui T. and Mori K., 'YAG laser lap welding of Zn-coated steels and monitoring of reflected beam', *IIW*, Bucharest, 2003 IIW Doc. IV-839-03
48. Goebels D., Kielwasser M. and Fabbro R., 'Improvement of laser welding of Zn-coated steel and aluminum alloys thin sheets using shaped laser intensity distribution, *Cong. Proc. of 22nd ICALEO 2003 (Laser Materials Processing Conf.)*, Jacksonville, LIA, 2003 (CD)
49. Gu H., 'A new method of laser lap welding of zinc-coated steel sheet', *Proc. ICALEO 2000*, Dearborn, LIA, 2000 **89** Section C 1–6
50. Dasgupta A., Mazumder J. and Bembenek M., 'Alloying based laser welding of galvanized steel', *Proc. ICALEO 2000*, Dearborn, LIA, 2000 **91** Section A 38–45
51. Ono M., Shinbo Y., Yoshitake A. and Ohmura A., 'Welding properties of thin steel sheets by laser-arc hybrid welding – laser focused arc welding', *Proc. of SPIE (First Int. Sym. on High Power Laser Macroprocessing)*, Osaka, JLPS, 2002 **4831** 96–100
52. Gu H. and Mueller R., 'Hybrid welding of galvanized steel sheet', *Proc. of ICALEO 2001 (Laser Materials Processing Conference)*, Jacksonville, LIA, 2001 Session A A304 (CD)
53. Todate A., Ueno Y., Katsuki M., Katayama S. and Matsunawa A., 'YAG laser weldability of carbon steel under CO_2 gas shielding', *Reprint of High Energy Beam Processing Committee*, Japan Welding Society (JWS), Tokyo, 2000 EBW-361-00 (in Japanese)
54. Katayama S., Yoshida D. and Matsunwa A., 'Assessment of YAG and CO_2 laser weldability in nitrogen shielding gas', *Proc. ICALEO 2000*, Dearborn, LIA, 2000 **89** Section C 42–51
55. Katayama S. and Matsunawa A., 'Solidification microstructure of laser welded stainless steels', *Proc. of ICALEO '84*, Boston, LIA, 1984 **44** 60–7
56. Katayama S., Iamboliev T. and Matsunawa A., 'Formation mechanism of rapidly quenched microstructure of laser weld metals in austenitic stainless steels', *5th Int. Conf. on Trends in Welding Research*, Georgia, ASM Int., 1998 93–8
57. Hiramoto S. and Ohmina M., 'CO_2 laser welding of aluminum in active gas shielding', *Journal of High Temperature Society*, HTSJ, 1990, **16**(3) 141–8 (in Japanese)
58. Takahashi K., Kumagai M., Katayama S. and Matsunawa A., 'Investigation of high-speed CO_2 laser welding of thin aluminum sheets', *J. of Light Metal Welding & Construction*, 2002 **40**(4) 177–81 (in Japanese)
59. Katayama S. and Lundin C.D., 'Laser welding of commercial aluminum alloys – laser weldability of aluminum alloys (Report II), *J. of Light Metal Welding & Construction*, JLWC, 1991 **29**(8) 349–60 (in Japanese)
60. Katayama S., Mizutani M. and Matsunawa A., 'Development of porosity prevention procedures during laser welding', *Proc. of SPIE (First Int. Sym. on High Power Laser Macroprocessing)*, Osaka, JLPS, 2002 **4831** 281–8

61. Katayama S., Kojima K., Kuroda S. and Matsunawa A., 'CO_2 laser weldability of aluminum alloys (Report 4), Effect of welding defects on mechanical properties, deformation and fracture of laser welds, *J. of Light Metal Welding & Construction*, JLWC, 1999 **37**(3) 95–103 (in Japanese)
62. Yamaoka H., 'Microstructural control of laser welded aluminum alloy', *J. of Light Metal Welding & Construction*, JLWC, 2001 **39**(5) 209–15 (in Japanese)
63. Schubert E., Zerner I. and Sepold G., 'No possibilities for joining by using high power diode lasers', *Proc. ICALEO '98*, Orlando, 1998 **85** Section G 111–20
64. Katayama S., Usui R. and Matsunawa A., 'YAG laser welding of steel to aluminum', *Proc. 5th Int. Conf. on Trends in Welding Research*, Georgia, ASM & AWS, 1998 467–72
65. Katayama S., Mizutani M. and Matsunawa A., 'Laser welding of aluminum and steel', Copenhagen, 2002 IIW, Doc. IV-814-02
66. Katayama S. and Mizutani M., 'Laser one-pass welding utilizing special lap- and butt-joint of dissimilar aluminum and steel', *Proc. of the 22rd ICALEO 2003*, Jacksonville, LIA, 2003, Section E 1401 (CD)
67. Katayama S., Morita M. and Matsunawa A., 'Laser joining of steel and magnesium alloy', *Proc. DIS '02 (Designing of Interfacial Structures in Advanced Materials and Their Joints)*, Osaka, 2002, 747–50
68. Jones I.A. and Hilton P.A., 'Use of infrared dyes for transmission laser welding of plastics', *Proc. of the 18th ICALEO '99*, San Diego, LIA, 1999 **87** Section B 71–9
69. Nakamura H., Terada M., Hirata M. and Sakai T., 'Application to the plastic parts welding of diode laser', *Proc. of 59th Laser Materials Processing Conference*, Nagoya, JLPS, 2003, **59** 1–7 (in Japanese)
70. Katayama S. and Mizutani M., 'Elucidation of laser welding phenomena and porosity formation mechanism', *Trans. of JWRI*, 2003 **32**(1) 67–9
71. Katayama S., Seto N., Mizutani M. and Matsunwa A., 'X-ray transmission in-situ observation of keyhole during laser spot welding and pulse-shaping for prevention of porosity', *Congress Proc. of ICALEO 2001 (Laser Materials Processing Conference)*, Jacksonville, LIA, 2001, Session C – Welding, 804
72. Katayama S., 'Laser Welding', *J. of Light Metal Welding & Construction*, JLWC, 2002 **40**(10) 476–87 (in Japanese)
73. Katayama S., Yoshida D., Yokoya S. and Matsunwa A., 'Development of tornado nozzle for reduction in porosity during laser welding of aluminum alloy', *Congress Proc. of ICALEO 2001 (Laser Materials Processing Conference)*, Jacksonville, 2001, LIA, C1701 (CD)
74. Petereit J., Abels P., Kaierle S., Kratzsch C. and Kreutz E.W., 'Failure recognition and online process control in laser beam welding', *Proc. of ICALEO 2002 (Laser Materials Processing Conference)*, Scottsdale, LIA, 2002 Session A – Welding (CD)
75. Kaierle S., Abels P., Kapper G., Kratzsch C., Michel J., Schulz W. and Poprawe R., 'State of the art and new advances in process control for laser materials processing', *Congress Proc. of ICALEO 2001 (Laser Materials Processing Conference)*, Jacksonville, LIA, 2001 E805 (CD)
76. Haferkamp H., Ostendorf A., Bunte J., Szinyur J., Hofemann M. and Cordini P., 'Increased seam quality for laser-GMA hybrid welding of zinc-coated steel', *Proc. of ICALEO 2002 (Laser Materials Processing Conference)*, Scottsdale, LIA, 2002 Session A – Welding (CD)
77. Schumacher J., Zerner I., Neye G. and Thormann K., 'Laser beam welding of aircraft fuselage panels', *Proc. of ICALEO 2002 (Laser Materials Processing Conference)*, Scottsdale, LIA, 2002 Section A – Welding (CD)

78. Travis D., Dearden G., Watkins K.G., Reutzel E.W., Martukanitz R.P. and Tressler J.F., 'Sensing for monitoring of the laser GMAW hybrid welding process', *Proc. of the 23rd ICALEO 2004*, San Francisco, LIA, 2004 (CD)
79. Muller-Borhanian J., Deininger C., Dausinger F.H. and Hugel H., 'Spatially resolved on-line monitoring during laser beam welding of steel and aluminum', *Proc. of the 23rd ICALEO 2004*, San Francisco, LIA, 2004 (CD)
80. Kogel-Hollacher M., Dietz C., *et al.*, 'Camera based process monitoring of the CO_2 and Nd:YAG laser welding experiences from applications in the automotive industry', *Proc. of the 23rd ICALEO 2004*, San Francisco, LIA, 2004 Sec. Sensing, Monitoring & Control 75–9 (CD)
81. Kawaguchi S., Mizutani M., Tarui T. and Katayama S., 'Correlation between in-process monitoring signal and welding phenomena in YAG laser welding of aluminum alloy', *Proc. of 62nd Laser Materials Processing Conference*, Osaka, JLPS, 2004 **62** 34–44 (in Japanese)
82. Kogel-Hollacher M., Jurca M., Dietz C., Janssen G. and Lozada E.F.D., 'Quality assurance in pulsed seam laser welding', *Proc. ICALEO '98*, Orlando, LIA 1998, **85** Section C 168–76
83. Miyamoto I. and Mori K., 'Development of in-process monitoring system for laser welding', *Proc. ICALEO '95*, San Diego, 1995, LIA, **80** 759–67
84. Olsen F.O., Jorgensen H. and Bagger C., 'Recent investigations in sensorics for adaptive control of laser cutting and welding', *Proc. LAMP '92*, Nagaoka, 1992, JLPS, **1** 405–14
85. Kawahito Y. and Katayama S., 'In-process monitoring and adaptive control for stable production of sound welds in laser micro-spot lap welding of aluminum alloy', *Journal of Laser Applications*, 2005 LIA, **17**(1) 30–7
86. Kawahito Y. and Katayama S., 'Adaptive control in laser micro-spot lap welding of aluminum alloy (Report I) – Adaptive control for fully-penetrated micro welding of thin sheets, *Journal of Japan Laser Processing Society*, 2004, JLPS, **11**(3) 154–9 (in Japanese)
87. Steen W.M. and Eboo M., 'Arc augmented laser beam welding, *Metal Construction*, 1979 **7**(7) 332–5
88. Beyer E., Dilthey U., Imhoff R., Majer C., Neuenhahn J. and Behler K., 'New aspects in laser welding with an increased efficiency', *Proc. of ICALEO '94*, Orlando, LIA, 1994 **79** 183–92
89. Ishide T., Tsubota S., Watanabe M. and Ueshiro K., 'Latest MIG, TIG arc-YAG laser hybrid welding system', *Journal of the Japan Welding Society*, 2003 **72** (1) 22–6
90. Ishide T., Tsubota S. and Watanabe M., 'Latest MIG, TIG arc-YAG laser hybrid welding systems for various welding products', *Proc. of SPIE (First Int. Sym. on High Power Laser Macroprocessing)*, Osaka, JLPS, 2002 **4831** 347–52
91. Petring D., Fuhrmann C., Wolf N. and Poprawe R., 'Investigation and applications of laser-arc hybrid welding from thin sheets up to heavy section components', *Proc. of the 22nd Int Congress on Applications of Lasers & Electro-Optics (ICALEO) 2003*, Jacksonville, LIA, 2003, Section A, 1–10 (CD:301)
92. Abe N., Kunugita Y. and Miyake S., 'The mechanism of high speed leading path laser-arc combination welding, *Proc. of ICALEO '98*, Orlando, LIA, 1998 **85** Section F, 37–45
93. Staufer H., 'Laser hybrid welding & laser brazing at VW and Audi', *Proc. of 6th High Energy Research Committee*, HiDEC-2003-01, 2003 1–10

94. Tsuek J. and Suban M., 'Hybrid welding with arc and laser beam', *Science and Technology of Welding and Joining*, 1999 **4**(5) 308–11
95. Beyer E., 'Laser technology for new markets – application highlights, *6th International Laser Marke Place 2003*, Anwendung im Dialog (2003) S. 5–15
96. Kutsuna M. and Chen L., 'Interaction of both plasma in CO_2 laser-MAG hybrid laser-hybrid welding of carbon steel, *IIW*, 2002, Doc. XII-1708–02
97. Schubert E., Wedel B. and Kohler G., 'Influence of the process parameters on the welding results of laser-GMA welding, *Proc. of ICALEO 2002 (Laser Materials Processing Conference)*, Scottsdale, LIA, 2002 Session A – Welding (CD)
98. Naito Y., Mizutani M. and Katayama S., 'Observation of keyhole behavior and melt flows during laser-arc hybrid welding', *Proc. of the 22nd Int Congress on Applications of Lasers & Electro-Optics (ICALEO) 2003*, Jacksonville, LIA, 2003, (CD: 1005)
99. Naito Y., Mizutani M., Katayama S. and Bang H.S., '*Proc. of the 23rd Int Congress on Applications of Lasers & Electro-Optics (ICALEO) 2004*, San Francisco, LIA, 2004 Hybrid laser welding (CD: 207) 41–9
100. Uchiumi S., Wang J.B., Katayama S., Mizutani M., Hongu T. and Fujii K., 'Penetration and welding phenomena in YAG laser-MIG hybrid welding of aluminum alloy', *Proc. of the 23rd Int Congress on Applications of Lasers & Electro-Optics (ICALEO) 2004*, San Francisco, LIA, 2004, Hybrid laser welding (CD: P530) 76–85
101. Tarui T., 'Trend of laser application for car body in European automotive industry', *Proc. of the 61st Laser Materials Processing Conf.*, Osaka, JLPS, 2004 **61** 152–7
102. Morita I., Yamaoka H., *et al.*, 'Study of underwater laser welding repair technology', *IIW*, Bucharest, 2003, IIW Doc.IV-846-03 & IIW Doc.XI-782–03
103. Ito H. and Okada N., 'Laser welding of electronics component', *Proc. of the 55th Laser Materials Processing Conf.*, Nagoya, JLPS, 2002 **55** 60–5

7
Electron beam welding

U. D I L T H E Y, RWTH-Aachen University, Germany

7.1 Introduction

The history of electron beam technology goes back to the year 1869 when Hittdorf discovered electron beams. Thomson, in 1897, found out about their negative electron charge. In 1905, Pirani was first to use electron beams for fusion tests with metals (Schiller *et al.*, 1977).

The principle of the technology is based on a beam of electrons that are accelarated by a high voltage and can so be used as a tool for treatment of materials such as in welding. Currently electron beam welding is firmly established in many manufacturing fields; especially in joining technology the electron beam has been generally accepted for its reliability and efficiency. The thickness of materials that can be joined ranges from thin plates of thicknesses of fractions of millimetres to thick plates of more than 150 mm in steel and over 300 mm in aluminium. Almost all electrically conductive materials can be welded and many such may be joined. The high power density ranges up to 10^8 W/cm^2 (typical of electron beam welding) and the connected depth-to-width ratio of the weld ranges up to 50:1. These figures allow a large variety of possible applications of the joining process. As the electron beam is electromagnetically deflectable and effectively lacking in mass, progress in the field of control techniques and increased processor performances have extended the variety of applications and these are still increasing. In some cases electron beam welding is carried out in a vacuum chamber (with high or low vacuum) and in others the electron beam can be applied under normal atmospheric conditions. The different variations in conditions for electron beam welding are shown in Fig. 7.1.

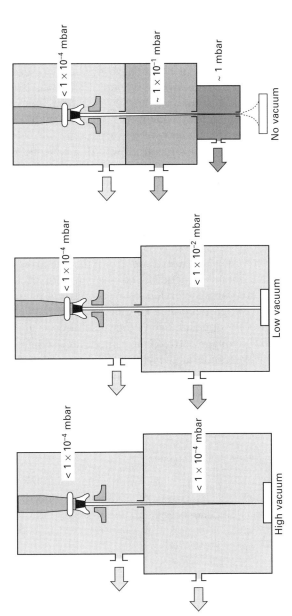

7.1 Varying pressure conditions under which electron beam welding can be carried out.

7.2 Basics of the process

7.2.1 Electron beam generation and guiding

Generation of the electron beam

Triode systems for beam generation are generally applied in modern electron welding machines as shown in Fig. 7.2. These systems are composed of anode, cathode and the control electrode (Wehnelt cylinder). The electrons that are necessary for beam generation are emitted from the cathode by thermionic emission. The cathode is made from a material such that the work that must be performed by the electron to leave is comparatively low. The cathode material must show a high electron emission rate, be resistant to high temperatures and guarantee a relatively long cathode life. Appropriate materials are tungsten and tantalum. The heating of the cathodes may be carried out directly or indirectly. Indirectly heated cathodes are heated by electron bombardment from an auxiliary cathode; current passes through those cathodes that are directly heated and therefore they are heated by Joule resistance. A high voltage electric field is applied which supplies the electrons with kinetic energy in order to emit them from the electron cloud and subsequently accelerate them. Depending on the strength of the applied voltage, the electrons may be accelerated up to two-thirds of the speed of light. An acceleration voltage generates an electric field between cathode and anode, situated directly opposite each other, (Schiller *et al.*, 1977). By the application of a control voltage between the cathode and a control electrode, (Fig. 7.2) a barrier field is generated in the triode system that forces the emitted electrons back to the cathode. Thus the beam current is controlled by alterations of the control voltage because through its decrease more electrons pass the barrier field towards the anode. Owing to its particular shape, similar to a concave mirror, the control electrode affects the electrostatic focusing of

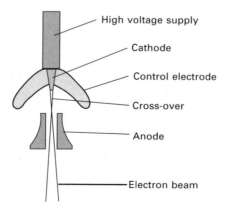

7.2 Triode system for beam generation.

the electron beam. After passing the anode the electrons have achieved their final speed and the electron beam is focused and deflected by means of electromagnetic focusing lenses. The focusing effect leads to the constriction of the electron beam, the so-called cross-over.

Beam manipulation

The electron beam diverges slightly after passing the pierced anode and is then focused to a spot diameter of between 0.1 and 1.0 mm by a beam manipulation system to reach the necessary power density of 10^6 to 10^7 W/cm^2. The beam is first guided through the alignment coil onto the optical axis of the focusing objectives. One or several electromagnetic lenses direct the beam onto the workpiece inside the vacuum chamber. Deflection coils that are positioned at various parts of the electron beam generator assist in the deflection or oscillating motions of the electron beam. A diagrammatic representation of an electron beam welding machine is depicted in Fig. 7.3.

7.2.2 Deep penetration effect

When the electrons strike the surface of the workpiece their kinetic energy is converted into thermal energy. Although the electron mass is very low (approximately 9.1×10^{-28} g) electrons have a high electric voltage potential which, at an accelerating voltage of 150 kV, allows electron acceleration up to a speed of approximately 2×10^8 m/s. Not all beam electrons penetrate the workpiece and release their energy to the material. Some of the striking electrons are emitted in other forms: back-scattered electrons, thermal radiation, secondary electrons or X-ray radiation as shown in Fig. 7.4.

Because of their low mass, the electrons that penetrate the material do so to only very shallow depths (of up to 150 μm), another process is needed in order to obtain large weld depths, the so-called deep-penetration effect. The material is melted and vaporised in the centre of the beam and this happens so quickly that the heat dissipation into the cold material has almost no effect. The resulting vapour is superheated to temperatures of above approximately 2700 K. The vapour pressure is sufficiently high to press the molten metal upwards and to the sides. A cavity develops where the electron contacts the yet unvaporised metal and heats this further. This leads to a vapour cavity which in its core consists of superheated vapour and is surrounded by a shell of fluid metal. This effect is maintained as long as the pressure from the developing vapour cavity and the surface tension of the molten pool are in equilibrium. The diameter of the vapour cavity corresponds approximately with the electron beam diameter. With a sufficiently high energy supply, the developing cavity penetrates through the entire workpiece (Schultz, 2000). The relative motion between workpiece and electron beam causes the material

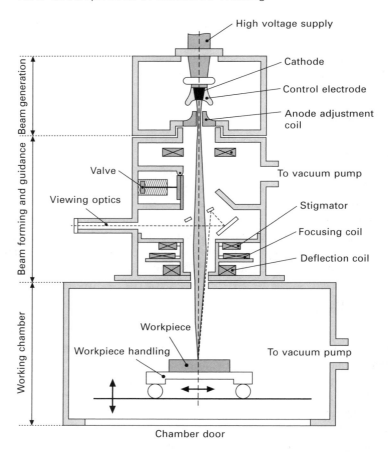

7.3 Diagrammatic representation of an electron beam welding machine.

which has been molten at the front of the electron beam to flow around the cavity and to solidify at the rear. The formation of the vapour cavity is depicted in Fig. 7.5 (Schiller *et al.*, 1977; Schultz, 2000).

The pressure and temperature conditions inside the cavity are subject to dynamic changes over time. Under the influence of the constantly changing geometry of the vapour cavity, welding faults such as shrinkage cavities may occur when the welding parameters have been chosen unsuitably. It is possible to avoid these faults by a suitable choice of welding parameters and, in particular, by the selection of suitable oscillation characteristics; examples are circular, sine, rectangular and triangular functions.

Electron beam welding 203

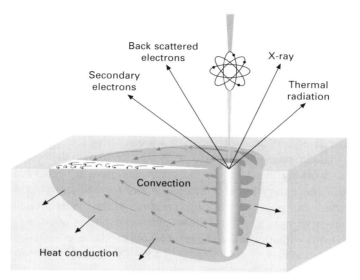

7.4 Fate of the electrons on meeting the workpiece.

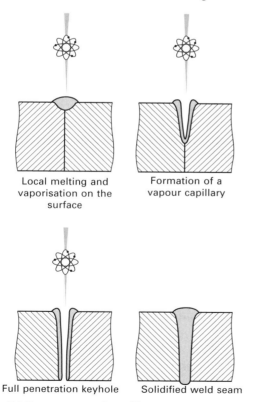

7.5 Deep penetration effect.

7.2.3 Machine components

The electron beam welding machine consists of a great many individual components. The basic component of the machine is the electron beam generator where the electron beam is generated in high vacuum, influenced by electromagnetic deflection coils and then focused onto the workpiece in the vacuum chamber (see Section 7.2.1). An electron beam in air diverges strongly through collision with air molecules and thus loses power, so welding is generally carried out in a low or high vacuum inside a vacuum chamber. Different vacuum pumps are used to generate a vacuum in the beam generator and in the working chamber. In the beam generator a high vacuum ($p < 10^{-5}$ mbar) is necessary both for insulation and for oxidation circumvention of the cathode but possible working pressures in the vacuum chamber vary between high vacuum ($p < 10^{-4}$ mbar) and atmospheric pressure. Collision of the electrons with any residual gas molecules and consequent scattering of the electron beam is obviously lowest in a high vacuum. The beam diameter of the focused electron beam is at a minimum in high vacuum and therefore the power density in the beam is at its maximum.

A shut-off valve positioned between the electron beam generator and the working chamber allows the presence of a vacuum in the beam generator area even when the working chamber is at atmospheric pressure. Other necessary features are a high voltage supply, controls for this supply, vacuum pumps, the numerical control (NC) of the work table and operator areas. The equipment is controlled at the operator console where all relevant process parameters are set and monitored. In modern equipment, parameter selection and control may be carried out externally by a computer and corresponding software. For the determination of optimum welding parameters for process control and also for the adjustment of the electron beam on the workpiece, viewing optic systems are necessary. Simple light-optical viewing optics with a telescope or a camera system and monitor which partially represent the magnified section, or electron-optical systems are used. In these systems the electron beam scans the workpiece surface with a very low power without melting it. The back-scattered secondary electrons show, as in scanning electron microscopy (SEM), an image of the workpiece surface. Figure 7.6 shows the electron beam welding machine and peripheral equipment.

Electron beam welding machines may be classified according to the quality of the vacuum, the machine concept and the height of the maximum acceleration voltage. The acceleration voltage exerts substantial influence on the achievable welding results – the higher the acceleration voltage, the lower the beam focus diameter of the focused beam at an equal beam power. Therefore, with a high acceleration voltage, the maximum achievable welding depth increases as does the ratio between depth and width of the beam geometry. However, a disadvantage of increasing acceleration voltage is the exponentially increasing

Electron beam welding 205

7.6 Electron beam welding machine and peripheral equipment.

X-ray radiation as well as increased sensitivity to flash-over voltages. In production systems a distinction is made between high-voltage machines with acceleration voltages of between 120 kV and 180 kV and low-voltage machines with acceleration voltages of maximum 60 kV. Beam powers of up to 200 kW are used.

7.2.4 Potential of fast beam controls

The electron beam is a welding tool with virtually no mass, which is deflectable, non-contacting and almost inertia-free. It is therefore possible to oscillate the beam with extremely high frequency and by applying a control voltage the beam may be switched off between the individual oscillations. With this technique the electron beam skips between several positions with a frequency so high that the metallurgical influence on the structure is carried out at different points simultaneously, due to the thermal inertia of the structure. Through recent developments in the field of beam deflection it is now possible to vary the focus position and the beam power between the individual oscillations and the beam can be controlled in such way that up to five electron beams simultaneously process the material as shown in Fig. 7.7. This technique offers considerable potential for many applications.

This technique can be most easily applied by forcing the beam to skip between two or more positions, thus producing, at a simultaneous movement

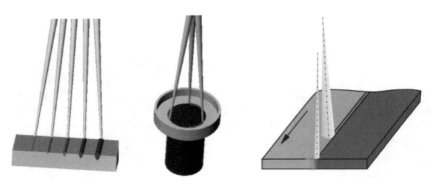

7.7 Multi-beam technique.

of the workpiece, two or more welds. The technique has been used for several years to join saw bands for band circular saws in conveyor units. These saw bands consist of a ductile backing layer in the middle of two hardened boundary layers, i.e. two parallel welds are necessary. A further interesting field of application for multi-beam technology is the welding of axis-parallel, rotationally symmetrical bodies. As the material melts simultaneously at several points of the axis-symmetrical weld and solidifies subsequently, the shrinkage stresses also occur simultaneously and symmetrically thus avoiding disalignment of the axes. This means that the often costly and labour-consuming press fits for centring and avoiding the disalignment of axes may be dispensed with. Another application of this fast beam deflection is the joining of material combinations. The multi-beam technique allows, by varying holding times at different points, the supply of one of the joining members at the welding point with a significantly higher energy than is supplied to the second member. For example, one joining member may be molten while the other one is simply heated (diffusion welding). In this way, it is possible to join materials that do not show complete solid solubility. Without the multi-beam technique, there is only a narrow range where the beam impact point opposite the joint groove can be used to apply different energy levels to the joining members. Because this variation demands extremely precise positioning of the beam it is difficult to reproduce.

7.3 Electron beam welding machines

Apart from the further development of beam generators, adaptation of the equipment to varying demands is of considerable importance to the industrial applications of electron beam welding. For example, the evacuation times in the vacuum chambers may be reduced so that the welding downtimes do not deter the use of electron beam technology. Different working chamber systems are currently available to equip electron beam welding machines.

7.3.1 Chamber machines

The most flexible variant is the universal working chamber where the workpiece is moved in two or three directions. Revolving devices with horizontal or vertical axes of rotation can be used instead of an NC coordinate table. Typical chamber sizes are from $0.1\,m^3$ to $20\,m^3$ and some machines can have a chamber volume of up to $3500\,m^3$. However when using vacuum chambers there can be comparatively high downtimes because such procedures as 'clamping the tool', 'entering the recipient', 'evacuation', 'welding', 'airing', and 'workpiece release' must be carried out one after another. Figure 7.8 shows a typical $2.5\,m^3$ chamber with an x-y-coordinate table.

In order to fulfil the demand for shorter cycle times, machine systems such as double chamber machines, lock chamber machines, cycle system machines and conveyor machines have been developed.

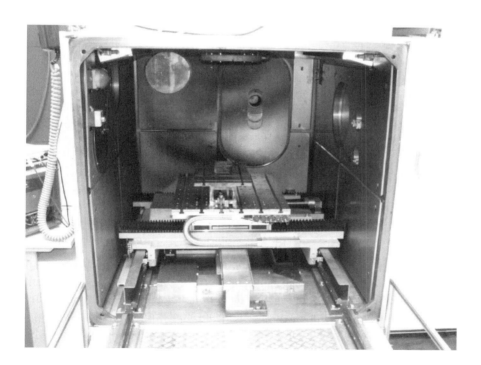

7.8 $2.5\,m^3$ universal vacuum chamber.

7.3.2 Double chamber and lock chamber machines

Double chamber machines have two working chambers that are placed side by side. The beam generator is either moved between the two chambers or the beam is deflected to one chamber at a time. Thus welding may be carried out in one chamber while the other chamber is loaded or discharged as well as evacuated. If the welding time exceeds the workpiece change and evacuation time, the capacity of both chambers is fully utilised. A disadvantage of these machines is that both chambers have to be equipped with separate movement devices and pumping units. Figure 7.9, left, shows one of the variations of a double chamber machine.

The lock welding machine is illustrated in Fig. 7.9, right. A high vacuum is permanently maintained in the chamber where the welding is carried out. Manipulation devices pass the workpieces through one or two prechambers. The machines have a position for loading and discharging, a lock for airing and de-aerating and a welding lock (von Dobeneck *et al.*, 2000). Figure 7.9 shows one type of a lock welding machine system.

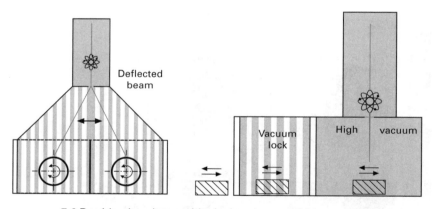

7.9 Double chamber and lock chamber machine.

7.3.3 Cycle system machines and conveyor machines

Cycle system machines are suitable for welding similar parts with equal weld geometries and axial welds; an example is shown in Fig. 7.10. Underneath the chamber is fixed a rotating jig, usually small-volume and demanding only short evacuation times, with vertical, horizontal or swivelling axes; this jig is equipped with one or several loading stations. This means that loading and discharging as well as welding may be carried out at the same time. In new cycle system machines there is a low vacuum in the area where the jig rotates all the time. This leads to shorter evacutaion times from free atmosphere to low vacuum at the loading position and from low vacuum to high vacuum at the welding positions, as shown in Fig. 7.10, right.

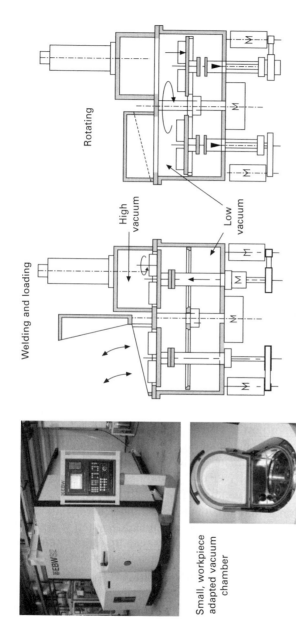

7.10 Cycle system machine (Source: PTR).

Conveyor machines are now available on the market especially for the manufacture of saw bands. This type of equipment is very productive but also very inflexible. Conveyor machines work on the the same operating principles as do lock welding machines. The workpiece is continuously transported over centring lips through the pressure locks into the working chamber and from there through a pressure lock. Any inevitable leakage is compensated by the vacuum technique.

7.4 Micro-electron beam welding

Owing to the very high functional integration density of the components and also to the great variety of materials, micro-systems and, in particular, hybrid micro-systems are making high demands on joining techniques. For example, the minute dimensions of the components and the frequent joining of dissimilar materials with differing thermophysical properties require joining methods that convey their energy selectively and locally in a minimum of space. Both laser beams and electron beams are suitable tools for this purpose. Moreover, processing in high vacuum meets the demands of a high-purity environment.

7.4.1 Technology

Electron beam welding machines from the macro-range cannot be used for micro-components. This is because their beam powers lie between 100 W and several kW which are too high for the welding of microfine components. The use of a scanning electron microscope (SEM) as a welding tool seems much more promising. This type of machine combines two basic functions, observation and welding, in just one piece of equipment. Calculations show that a maximum acceleration voltage of 30 kV and a probe current of 200 μA give a maximum beam power of 6 W in the beam generator. Power losses caused by screening through a diaphragm and scattering in the beam column reduce the power to approximately 3 to 4 W at the workpiece.

Figure 7.11 shows the beam path of the welding equipment in both operating conditions, observation and welding. The integration of both functions into one piece of equipment makes opposing demands on the technique. The observation and analysis of substrates require low energy input and high resolution. The high number of diaphragms that are small enough to screen edge electrons and the two condenser coils that reduce superfluous electrons in the beam cause extremely small beam radii. On the other hand, in most welding applications a higher power, by several orders of magnitude, is necessary. The technical modifications that have been carried out successfully are reversible and are primarily concerned with the electromagnetic components. Among such modifications are the removal of two diaphragms from the liner tube and the increase of the aperture diaphragm diameter as

Electron beam welding 211

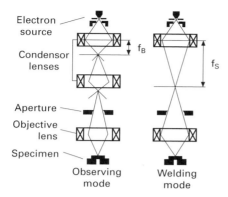

7.11 Micro-electron beam welding unit.

well as the switching off of the first condenser coil. Because these modifications are reversible, the existing equipment may be used both for observation and for welding (Carslaw and Jaeger, 1967).

In the practical applications of the micro-system technique, micro components are joined to one another in varying arrangements. The quality of a welded joint is strongly dependent on the adjustment precision of the components to be joined, among other factors such as joint preparation. For instance, a slight angle deviation of two components, joined with a square butt joint can lead to a gap no longer compensated by the minute beam diameter of a few micrometers. Faulty joints are a consequence. Owing to the excellent observation possibilities of SEM, the highly precise adjustment is carried out only in its working chamber. This means that the existing coordinate system in the interior of the compound chamber has, with the five-axes macro table, been completed to a second independent coordinate system. In addition, a subsequent correction of the component position without repeated withdrawal of the components from the compound chamber is possible.

The adjustment device is composed of three linear adjusters, two tilting axles and one control unit. The vacuum suitability of the components allows the maintenance of the necessary vacuum and the reliable operation of the motors. Self-locking of the gears allows the controls to be switched off after reaching the desired position and leaves the electron beam uninfluenced by electromagnetic fields.

When choosing and integrating the second adjustment unit into the vacuum chamber of the SEM, several conditions must be considered:

- Vacuum suitability of the mechanical and the driving components;
- High plane-parallelism of adapter plates;
- Circumvention of collision with wall, electron gun and detectors during operation;
- Centre position of all axes below the exit outlet of the electron beam;

212 New developments in advanced welding

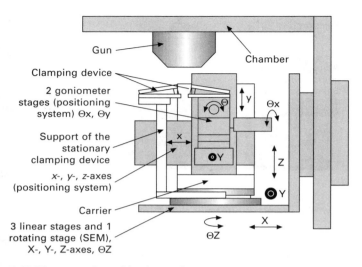

7.12 Diagram of partitioning unit.

- Z-position of the joining plane must be located in the region of the working distance of the electron beam;
- Vacuum pipe for the electrical drive of the motors.

The system design allows one joining component to be moved by the five-axes-positioner independently of the second joining component. Figure 7.12 shows the positioning unit and its diagrammatic representation after installation in the SEM.

7.4.2 Micro-welding process

Process sequence

Figure 7.13 shows the chronological sequence of the welding process. As a first step, the components to be joined are adjusted exactly in relation to each other and the electron beam is positioned on the joint. There is then a changeover to the welding mode so that the actual welding operation can start; however, on-line process observation is not yet possible. After the welding process is completed, the joining point may, after changeover to the observation mode, be subject to further analyses or measurements. In practice all welding sequences correspond with this step sequence.

Process variations

In practice, there is the choice of several welding methods, (Fig. 7.14). They basically differ in the type of beam manipulation on the substrate. During single scanning the electron beam is guided once over the welding zone with

Electron beam welding 213

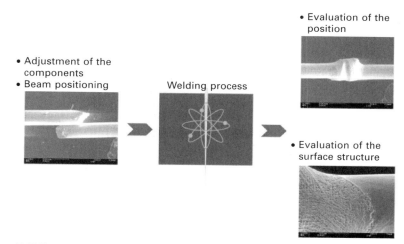

7.13 Process sequence in micro-electron beam welding.

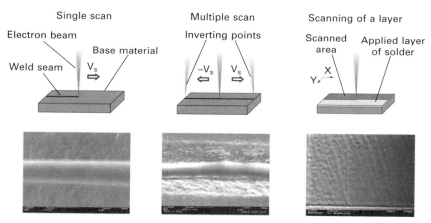

7.14 Process variations.

a fixed welding speed; during multiple scanning it oscillates for a certain period of time with a preselected deflection frequency. The choice of method depends on the joining task. Single scanning for the joining of foils leads to very clean, sharply delimited weld edges without weld notches; multiple scanning, however, shows weld edge regions that are remarkably elaborated. The beam energy, absorbed over a longer period of time, leads to the partial evaporation of the metal. Energy input can be varied further by the scanning of a larger substrate surface on a lateral level. Here the electron beam is applied as the heat source for a soldering process. In a first step, low-melting soldering materials are applied on the micro-components that are to be joined. The electron beam leads, through its thermal influence, to the development of intermediate constituent phases inside the soldering materials and these in turn result in the joining of the components. In particular, materials with good thermal conductivity, or non-metals may, under certain prerequisites, be joined by means of micro-electron beam soldering.

7.4.3 Examples of joining

Knowledge about the methodology and equipment of electron beam welding has primarily been applied to wire joints and metal foils. Additionally, weld-in tests have been carried out to investigate the behaviour of silicon and plastics. Joining thermoelements made from NiCr–Ni wire combinations with a wire of diameter 70 μm each allows almost globular beads for temperature measurement in the micro range. In the field of plastics, polyethylene showed, after prior gold plating, good welding results. By means of materials with favourable heat-conductive properties, such as copper or aluminium, micro-soldering with the electron beam as the heat source is examined using Cu–Sn soldering systems (Janssen, 1991).

7.4.4 Summary

The understanding of the processes concerning electron beam welding in micro-technology and the extension of the equipment technique are consolidated by further research work in the fields of beam characterisation, beam–material interactions, temperature measurements and the integration of image processing in the control of welding processes. An important future aspect of research is the investigation of welding tasks with optional weld seam geometries. The production of appropriate control for the beam deflection and of means of suitable heat control are the most important challenges.

7.5 Non-vacuum electron beam welding

In the early stages of electron beam technology development, research was carried out in Germany on methods to guide the beam from the vacuum environment of the beam generator to the atmosphere. This became the basis of the non-vacuum electron beam welding (NV-EBW) process. The substantial weld depths which characterise vacuum electron beam welding (as a result of the power density of the beam) are not achievable with the NV-EBW method. The strong point of NV-EBW lies mainly in high-speed production. Achievable welding speeds reach up to 60 m/min when welding aluminium sheets and up to 25 m/min when welding steel plates.

Industrial researchers in the United States recognised the potential of NV-EBW early and advanced it to a further universally applicable joining method. For better energy coupling to the workpiece, beam generators with an accelerating voltage of 175 kV are used (Dilthey and Behr, 2000; Draugelates *et al.*, 2000). While the NV-EBW technology was taken up immediately in the United States and applied successfully, the method attracted less attention in other countries. The car industry has recently, due to the high efficiency of the electron beam and also to the high achievable welding speed, become interested in this method.

7.5.1 Technology

Vacuum-related restrictions can be overcome by guiding the vacuum-generated electron beam that has exited the beam generator to the atmosphere, over a multi-stage orifice assembly and nozzle system. The pressure chambers have a correspondingly high pressure (10^{-2} up to 1 mbar). They are connected after the beam generator chamber (vacuum 10^{-4} mbar), are evacuated separately and are separated from each other by pressure nozzles. The electron beam is focused on the exit nozzle which has an inner diameter of 1–2 mm. After its exit to the atmosphere the electron beam collides with the air molecules and expands, Fig. 7.15 (Schultz, 2000; Behr, 2003).

The scattering of the electron beam is reduced by increasing the accelerating voltage and by the application of helium as a working gas. For the effective utilisation of the helium gas flow, a coaxial gas jet is applied at the beam exit outlet. The effect of the electron scattering at the gas molecules is additionally attenuated by the strong heating of the gas in the electron path. This reduces

7.15 Photograph of the ISF-non-vacuum nozzle system.

gas density and decreases scattering (Dilthey and Behr, 2000; Draugelates *et al.*, 2000; Schultz, 2000; Behr, 2003).

As in laser beam welding, plasma formation occurs in NV-EBW. However, in NV-EBW the corpuscular character of the beam causes the plasma to be 'transparent' to the electron beam. For this reason the reflection properties of the material to be processed do not play a role during beam coupling (Behr, 2003).

After their exit from the orifice assembly, the electrons of the focused beam impinge on the material surface with a high speed and transmit their kinetic energy to the material lattice. This causes an increase in the kinetic energy of the lattice atoms and, when the power density in the focal spot is sufficiently high, the material temperature rises and even exceeds the boiling point of the material to be welded. However, not all beam electrons participate in the conversion of kinetic energy to thermal energy. The collision of the accelerated electrons with the mass particles of the air and of the joining materials causes, depending on the acceleration voltage and the density of the material, X-ray radiation. This radiation must be shielded. In addition, the ionisation of the air causes the generation of ozone, which must be neutralised. A radiation-proof working chamber should be designed with suitable materials and dimensions. Limits to the component size are, therefore, not much higher in NV-EBW than they are in Nd:YAG laser beam welding.

During welding, the power density may be varied either via the beam current or via the working distance and/or the gas atmosphere. However, the power density is not the only decisive factor for the welding result: for each material, several parameters must be taken into account, alone or in combination. Such parameters include working distance, welding speed, beam current, electron beam incident angle, possible gas supply and possible wire supply. For a successful application of NV-EBW it is necessary to know what parameters affect the process and how their interactions influence it (Fig. 7.16). Only then can the potential user consider the limits and possibilities of any given method for a particular joining task.

NV-EBW can allow significant increases in productivity. In the course of several years' research and development acitivities in the field of NV-EBW it has been shown several times over that the characteristic properties of this welding method are hardly known in industry. This lack of awareness complicates the introduction of NV-EBW (where the disadvantages are known) as it does for many manufacturing processes. The development of X-ray radiation can initially limit the introduction of a new welding method. This criterion seems to characterise NV-EBW as a particularly hazardous process. However, there are strict radiation protection guidelines and when these are followed electron beam equipment will be radiation tight and fitted with several safety measures. Accidents involving radioactive contamination from the NV-EBW machines have not been heard of up to the time of writing.

Electron beam welding 217

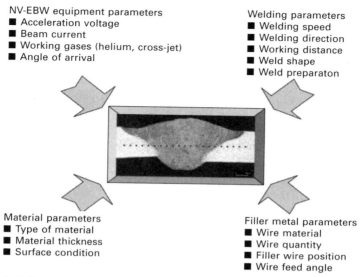

7.16 Process parameters in NV-EBW.

7.5.2 Applications of NV-EBW

A high-power and out-of-vacuum electron beam is the ideal tool for welding conventionally manufactured sheets and sheet metal parts. The upper bead of the weld is similar to that of an arc weld and thus cannot be compared with the typically narrow deep geometry of vacuum electron-beam welded joints. The method is characterised by high energy efficiency; its high available beam power yields a high power density even when the beam is expanded and allows high welding speeds (Fig. 7.17).

The application of NV-EBW is particularly recommend when high weld speeds and short cycle times with smaller weld depths are required at the same time. The main application field is thin sheet welding (thicknesses from 0.5 mm up to 10 mm).

A further field of NV-EBW is the welding of tailored blanks which is today applied extensively in the car industry and in terotechnology. The made-to-measure plates are produced through joining different plates with varying thickness, qualities and surface coatings to suit their future loads during application. The plates thus produced may afterwards be formed into a component and then welded to a shell. Manufacturing of tailored blanks shows the advantages of NV-EBW: cost saving through the reduction of parts and tools and through a substantial reduction of materials and assembly time. The component and/or the car as the final product may therefore be reduced in weight and its fuel consumption will be lower.

A classical application field of NV-EBW is the welding of components

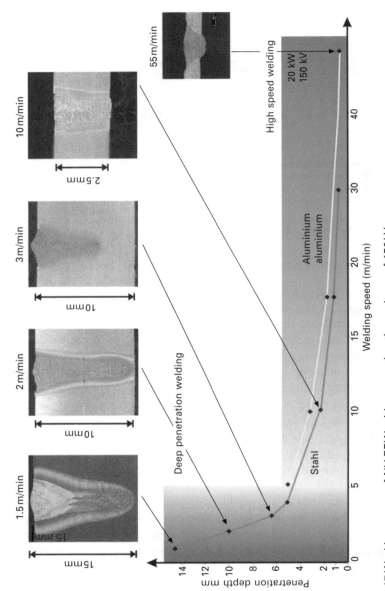

7.17 Working range of NV-EBW with an acceleration voltage of 150 kV.

Electron beam welding 219

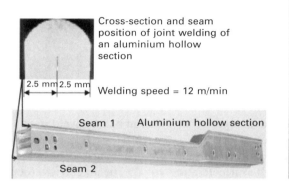

7.18 NV-EBW equipment for manufacturing Al-hollow sections.

where several plates which form a flange weld are joined, Fig. 7.18. Non-vacuum electron beam welding is very suitable for this application. The broad beam fuses several plate edges simultaneously which leads to a gas-tight and even joint. The flange weld and the lap joint are particularly suitable for components where, after a very rough weld preparation, the desired result is a gas- and liquid-tight weld.

The materials that have been tested up to now with the NV-EBW method are uncoated and coated steels, light metals such as aluminium and magnesium and non-ferrous metals, such as brass or copper. Material combinations, like for instance, that of steel and copper may also be realised with results comparable to vacuum EBW, without, however, achieving the higher weld depths of vacuum EBW. Supplementary application tests on the use of filler wire in NV-EBW have been carried out.

7.5.3 Summary

Non-vacuum electron beam welding is an efficient, reliable and well-known tool for material processing in welding. NV-EBW is, compared with other fusion welding methods, characterised by a lower thermal load on the point of effect and by high process speeds. Non-vacuum electron beam welding has the advantage that the process is not dependent on a vacuum chamber. The application of the NV-EBW method is recommended whenever high welding speeds and short cycle times with smaller weld depths are required at the same time. Non-vacuum electron beam welding directly competes with laser beam welding with a beam power of up to 20 kW. With an equipment efficiency of approximately 60% the non-vacuum electron beam is clearly the more efficient tool.

As a joining technique, NV-EBW is gaining in importance and NV-EBW may, in future, provide significant competitive advantages through its specific properties.

7.6 Quality assurance

7.6.1 Beam measurement

For a full exploitation of the advantages of the electron beam a welding, tool knowledge of beam properties is necessary. The processes that occur in electron beam welding are very complex and are characterised by a great many different parameters such as accelerating voltage, beam current, focus position and power density distribution. To determine the beam parameters and to facilitate the parameter transfer between different electron beam units a number of beam-diagnostic systems, which apply different measurement principles, are currently under development (Elmer and Teruya, 2001; Akopiants, 2002; Bach et al., 2002). The DIABEAM (Dilthey et al., 1992, 1997, 2001) has been developed in the ISF-Welding Institute, RWTH-Aachen University. This system may be employed in almost all existing electron beam units and allows signal acquisition of the electron beam up to a power of 100 kW. The DIABEAM measurement system was developed for easy determination of the focus position with slit measuring or a rotating wire, eliminating the need for complex and cost-intensive welding tests. The measurement and the three-dimensional display of the power density distribution across the beam diameter can be made by means of the apertured diaphragm. The other purpose of beam diagnostics, by in-time identification of variations of beam characteristics, is the prevention of negative influence, caused for instance by cathode adjustment, cathode distortion or variation of the vacuum level, on the welding result. As a result of the three measuring processes (hole, slot and rotating wire), the DIABEAM beam diagnosis system is suitable for a broad range of applications, especially for analysing and quality assurance of the beam.

As part of a common European research project, 12 different electron beam welding machines all over Europe have been measured (Dilthey and Weiser, 1994). A general dependence between electron beam parameters and welding results has been established and the power density, necessary for the formation of a vapour capillary, was experimentally determined.

The DIABEAM system deflects the measured beam over a combined double slit-hole sensor using a deflection unit or measures the undeflected beam with a newly developed rotating wire sensor device. The sensor transmits the signals via an amplifier integrated into the sensor case to a transient storage card (maximum scanning rate 40 MHz). In the measurement mode the beam is deflected with a very high deflection velocity of 200–900 m/s, depending on the beam power, by a computer controlled function generator with amplifier, which prevents destruction of the sensor.

The deflection is effected by a special coil, which enables a maximum deflection angle of ± 8. Each time the beam passes over the sensor, the measured data are read in by the transient storage card and displayed online on the computer monitor making correction adjustments possible.

To understand the results of measurements delivered by the DIABEAM system it is necessary to know the different definitions of beam diameters used by DIABEAM. There are two alternative diameter definitions that use interpretation of the width of the measured electrical signal or statistical evaluation of obtained power density distribution for the correspondent diameter definition. Figure 7.19 shows the layout of the DIABEAM system adapted to an EB welding equipment measured signal interpretation. As the electron density in the beam is proportional to the measured signal amplitude the most straightforward way to define beam diameters is illustrated in Fig. 7.20(a) for the one-dimensional rotational symmetrical case.

The diameter is defined as the width of the distribution for a given fraction of the maximum signal amplitude. In a two-dimensional case the contour plot of the beam may be used for the diameter definition, Fig. 7.20. The contour lines correspond to a given fraction of the maximum signal amplitude. The beam area including all amplitudes which exceed a given amplitude fraction defines the contour area A_x. The diameter d_x (where x is the fraction of the amplitude in %) is defined as the diameter of the corresponding circle having the area A_x:

$$d_x = 2\sqrt{\frac{A_x}{\pi}} \qquad [7.1]$$

Interpretation of power density distribution

In the case of a two-dimensional measurement (hole measurement) there is another possible way to define beam diameters. Instead of using the decrease of the signal amplitude, one can use a definite fraction of the power density integral over an area to the total beam power. In other words, the power density flux which crosses the beam area should constitute X % of the total beam power. Figure 7.20(b), illustrates the above definition in a one-dimensional projection of a measured two-dimensional power density distribution.

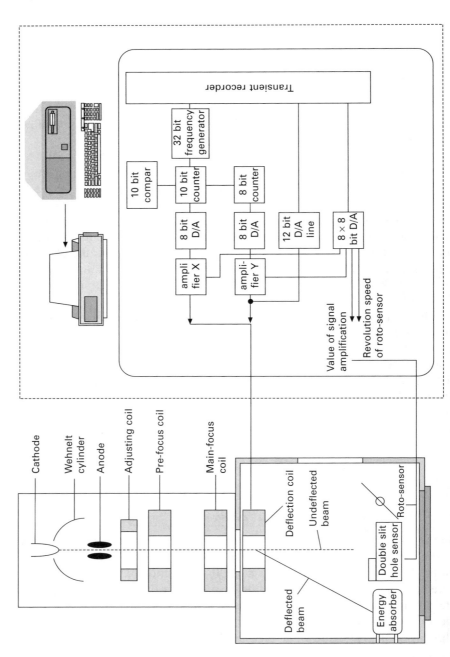

7.19 Layout of the DIABEAM system adapted to an electron beam welding equipment measured signal interpretation.

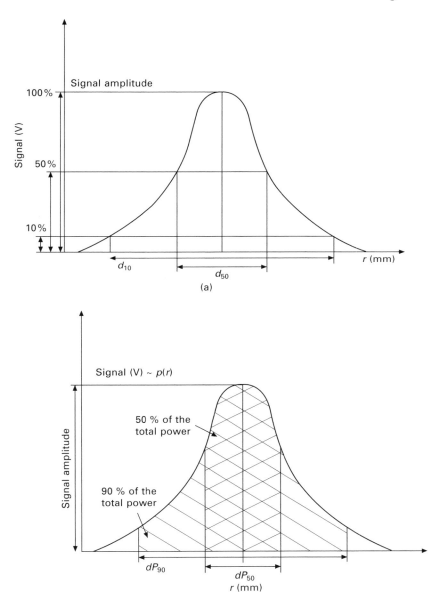

7.20 Two alternative methods of diameter definition.

In the simplified rotational symmetrical case it can be written:

$$P_x = \int_0^{\frac{d_{px}}{2}} p(r) 2\pi \, dr \qquad [7.2]$$

where $p(r)$ is the two-dimensional power distribution function, d_{PX} is the beam diameter and P_X is the beam power included in the area with diameter d_{PX}.

The diameter d_{PX} is once again defined as the diameter of a circle with equivalent area A_{PX}.

In the general case one can write:

$$\frac{P_x}{P_{total}} = X = \int_{A_{px}} p(x,y)\,dA \qquad [7.3]$$

and

$$d_{px} = 2\sqrt{\frac{A_{PX}}{\pi}} \qquad [7.4]$$

In DIABEAM the diameter based on the power definition is called the d_{PX} diameter (e.g. d_{P50}), where X is the fraction of total power in %. For all DIABEAM measurements five different contour values are evaluated according to one of the two described methods.

7.6.2 Sensor systems

Scanning can alternatively be carried out via a slit- or apertured diaphragm or via a rotating tungsten wire. Slit measurement with slit widths of 20 µm is a comparatively fast and simple determination of a signal which is proportional to the beam intensity and also the determination of beam diameter. The principle of slit measurement with the appropriate voltage signal over the beam cross-section is depicted in Fig. 7.21.

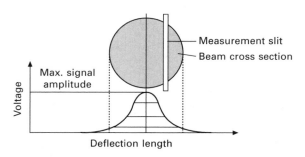

7.21 Principle of slit measurement.

With the slit measuring process, the core and edge areas of the beam can be compared by means of five different selectable diameters. Measurement of the beam diameter under varying working distances enables the caustic curve of the beam to be determined and displayed. In this way, the beam aperture can be determined precisely, thereby simplifying considerably the welder's selection of electrical and geometric parameters.

The application of a double-slit sensor enables online measurement of the deflection speed and increases the precision of the measurement. Here, the beam is, at the start of the measurement, deflected from its neutral position transversely over both slit sensors; there is no deflection in the longitudinal direction. The deflection speed is determined by the measured difference in time of the signals arriving at the first and at the second slit.

The apertured diaphragm measuring process enables detailed assessment of the electron beam. The high local and temporal resolution (up to 400 lines per mm and 40 M samples/s) of the measuring system enables the characteristics of the beam to be assessed with regard to changes in it in relation to certain parameters. These parameters may be electrical or non-electrical and include cathode adjustment, cathode deformation, vacuum and working distance. The apertured diaphragm measurement is carried out by deflecting the beam via the hole on a line-by-line basis. During fast deflection, recording of the signal (i.e. the current of electrons conducted from the hole) by a transient storage card is initiated by the DIABEAM hardware, Fig. 7.22(a). The beam is, in accordance with the parallel arrow lines (shown in the figure), deflected over the sensor in the X-direction. Between the individual parallel scans a static deflection in the Y-direction is carried out in every case. In general, 50 passes are carried out. Thus the power density distribution in the beam may be drawn up in a three-dimensional representation, as shown in Fig. 7.22(b).

The rotating wire sensor is also used in beam diagnostics. This new measuring variation has been developed to investigate whether and to what degree metal ions influence the power density distribution. Here a rotating tungsten wire with a diameter between 0.1 mm and 5 mm and at a speed of

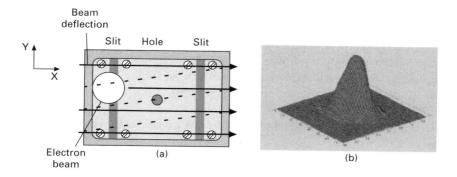

7.22 Type of beam deflection pattern for hole measurements (a) and corresponding 3D-representation (b).

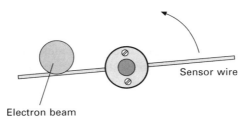

7.23 Principle of the rotating sensor.

up to $1.000\,\text{s}^{-1}$ moves through the beam; the principle is illustrated in Fig. 7.23. Beam deflection is unnecessary here. The tungsten wire is coupled to a solid copper plate in order to increase heat dissipation and the current derived from the wire measured in the form of a voltage signal. The measuring principle is similar to the slit measurement principle except that the diameter of the tungsten wire is smaller than is the diameter of the beam. In principle both methods (the rotating sensor and slit measurement) can be used to carry out the same type of beam diagnostics. An advantage of the rotating wire method is that it is possible to measure the non-vacuum electron beam; this rapidly spreads out in the atmosphere due to the scattering of the beam electrons on contact with air molecules (see Section 7.5.1).

7.7 Applications

Because of the great many materials that can be welded with the electron beam, such as tungsten, titanium, tantalum, copper, high-temperature steels, aluminium and gold as well as the large range of thicknesses that can be worked on, the method has a wide variety of applications. Such applications range between the micro-welding of sheets with thicknesses of less than 0.1 mm (here low and extremely precise heat input is important) and thick plate applications.

In heavy plate welding, the advantages of the deep penetration effect and the consequent joining of large cross-sections in one working step using a high welding speed, low heat input and small weld width become obvious. With modern welding equipment, wall thicknesses of more than 300 mm (aluminium alloys) and of more than 150 mm (low and high-alloy steel materials with length-to-width ratios of approximately 50:1) are joined quickly and precisely in one pass and without filler metal.

Listed below are some industrial applications where electron beam welding is an established tool.

- Reactor construction and chemical apparatus engineering: welding of high-alloy materials, welding of materials with high affinity for oxygen, production of fuel elements and of circumferential welds of thick-walled pressure vessels and pipes;

- Pipeline industry;
- Turbine manufacturing: production of guide blades and distributors;
- Aircraft construction: welding of structural/load-bearing parts made of titanium and aluminium alloys and of landing gears made of high strength steels;
- Automobile industry: welding of driving gears, pistons, valves, axle frames and steering columns;
- Electronics industry;
- Tool manufacturing, e.g. manufacturing of bimetal saw bands;
- Surface treatment;
- Material remelting;
- Electron beam drilling with up to 3000 drills per second (Dilthey, 1994).

Because of the numerous advantages (such as minimum workpiece heating by high power methods, small beam diameters and high welding speeds) electron beam welding is being increasingly applied in industrial practice. Low heat input also allows the welding of readily machined parts. The economic profitability of electron beam welding is partly due to the high welding speeds and therefore the short cycle times and partly due to the high quality. The clean vacuum process without the presence of oxygen together with constant process parameters make the weld seams easy to reproduce.

There are also some disadvantages. The workpieces to be welded must be electrically conductive and there is the risk of hardening and cracking due to high cooling rates; this restricts the range of materials that can be worked by the process. Investment costs are high because the beam deflection is carried out using magnetic fields and the whole process needs to be shielded because of the development of X-ray radiation during welding.

7.8   References

Akopiants K.S., (2002), 'System of diagnostics of electron beam in installations for electron beam welding', *The Paton Welding Journal*, **10**, 27–30

Bach F.W., *et al.* (2002), 'Non vacuum electron beam welding of light sheet metals and steel sheets', *IIW Document* Nr. IV-823-02

Behr W., (2003), *Elektronenstrahlschweißen an Atmosphäre*, Aachen, Shaker Verlag

Carslaw H.S. and Jaeger J.C., (1967), *Conduction of Heat in Solids*, Oxford, Clarendon Press

Dilthey U., (1994), *Schweißtechnische Fertigungsverfahren*, Düsseldorf, VDI-Verlag

Dilthey U. and Behr W., (2000), 'Elektronenstrahlschweißen an Atmosphäre', *Schweißen und Schneiden*, **8**, 461–85

Dilthey U. and Weiser J., (1994), 'Analysis of beam/workpiece interaction applied to electron beam welding for industrial application', *Final report: BREU-0134-CT90*

Dilthey U., Ahmadian M. and Weiser J., (1992), 'Strahlvermessungssystem zur Qualitätssicherung beim Elektronenstrahlschweißen', *Schweißen und Schneiden* **44** (4), 191–4

Dilthey U., Böhm S., Dobner M. and Träger G., (1997), 'Comparability and replication of the electron beam welding technology using new tools of the DIABEAM measurement device', *EBT '97: 5. Internat. Conf. on Electron Beam Technologies*, 76–83, Varna, Bulgaria

Dilthey U., Brandenburg A., Möller M. and Smolka G., (2000a), 'Joining of miniature components', *Welding and Cutting*, **52** (7), E143–E148

Dilthey U., Smolka G., Lugscheider E. and Lake M., (2000b), 'Electron-beam-induced phase generation at solder systems applied with high-performance cathode sputtering', *VTE*, **13** (1), E9–E15

Dilthey U., Goumeniouk A., Böhm S. and Welters T., (2001), 'Electron beam diagnostics: a new release of the DIABEAM system', *Vacuum*, **62**, 77–85

Draugelates U., Bouaifi B. and Ouaissa B., (2000), 'Hochgeschwindigkeits-Elektronenstrahlschweißen von Aluminiumlegierungen unter Atmosphärendruck', *Schweißen und Schneiden*, **52**, 333–8

Elmer J.W. and Teruya A.T., (2001), 'An enhanced Faraday cup for rapid determination of power density distribution in electron beams', *Welding Journal*, **80**, 288–95

Janssen W., (1991), *Verbesserung des Elektronenstrahlschweißens mit Hilfe der flexiblen Doppelfokussierung*, Aachen, VDI Verlag

Schultz H., (2000), *Elektronenstrahlschweißen*, Düsseldorf, Deutscher Verlag für Schweißtechnik (DVS)

Schiller S., et al. (1977), *Elektronenstrahltechnologie*, Stuttgart, Wissenschaftliche Verlagsgesellschaft

von Dobeneck D., Löwer T. and Adam V., (2000), *Elektronenstrahlschweißen – Das Verfahren und seine industrielle Anwendung für höchste Produktivität*, Verlag Moderne Industrie, Landsberg

8
Developments in explosion welding technology

J. BANKER, Dynamic Materials Corporation, USA

8.1 Introduction

Explosion welding technology (EXW) utilizes the energy of a detonating explosive to create conditions which result in welding between metal components. The technology is typically considered to be a cold welding, non-fusion process. For practical purposes, the process is typically limited to creating welds between the faces, or planar surfaces, of metal components. It is generally not applicable for production of traditional butt welds. It is most suitable for producing large area planar welds that are typical in clad metals. There are many good publications discussing EXW technology to which the reader is referred for an in-depth discussion; only primary basic process information, new developments and practical applications will be reported in this chapter.[1-6]

8.2 Capabilities and limitations

8.2.1 Applicable metals

EXW is a highly versatile technology from the materials applicability perspective. It is suitable for joining both metal combinations of similar composition and metal combinations of highly dissimilar composition. The latter has been the primary impetus for the commercial development of the technology. Today two of the primary applications of explosion welding are the production of reliable weld joints between aluminum and steel, and between titanium and steel. The success with these metal systems results from the absence of any significant bulk heating of the metals. The temperature–time conditions that cause the formation of brittle intermetallic compounds do not exist. Consequentially, explosion welding can be used to join almost any combination of metals. The factors limiting the suitability of EXW are primarily mechanical. During the EXW process, the metals are subject to high impact loading and significant cold deformation. A minimum ductility of 15% and a minimum fracture toughness of 50 J are generally considered the practical limits for successful EXW welds.

8.2.2 Sizes and thicknesses

EXW is ideally suited for making planar welds between metal plates or sheets. The thickness of the flyer plate, often called the cladder, can range from 0.1 mm to about 50 mm. For practical commercial reasons, costs are minimized when the flyer plate thickness is about 2–3 mm. The thickness of the base plate can range from 0.1 mm to over 1 m. In the case of the base metal, costs are minimized when the base plate thickness is around 12 mm.

The lateral size of clad plates, length and width, is primarily limited by the size of metal sheets or plates that are commercially available, not by the technical limits of the process. The commercial metal sizes available vary considerably between the different metal types. For most common commercial flyer metals, widths at 3 mm and less are limited to 1.2 m; thicker gages are commonly produced in widths of 2.5 m to 3.5 m dependent upon alloy type. For many metal types, two or more sheets can be edge butt seam welded using common fusion processes to increase plate size options. For example, for production of clad plates of 3 mm nickel alloy onto 25 mm thick steel, up to 4 m × 10 m plate size is not uncommon.

8.2.3 Product contour

Because of the direction of the jet in the EXW process, it cannot be used for welding onto three dimensional contoured surfaces. EXW is limited to cladding of flat plate surfaces or to cladding of concentric circular surfaces of straight bars, tubes or pipes.

8.2.4 Production facilities and commercialization

The amount of explosive required for production of most components is considerable. Although very small welds can be made under typical shop conditions, most explosion welds are produced under conditions where hundreds or thousands of kilograms of explosive can be detonated without damage to surroundings. Remote open air detonating sites or massive shooting chambers are commonly used. Restrictions on the commercial availability of the explosive products further limit broad applications of the process. At the time of writing there are about 30 to 40 companies commercially using EXW worldwide.

8.3 EXW history

For well over a century, there have been infrequent reports of metal pieces being 'welded' together during military detonation situations. During the late 1950s several institutions worldwide increased R&D activities in the realm of explosion metalworking. In 1960, DuPont filed the first internationally recognized patent on EXW.[7] During the ensuing 20 years there was extensive research concerning the technology. In 1962 DuPont commercialized the explosion cladding industry, with the first major application being production of tri-layer coinage for the US government. During the late 1960s DuPont codified the production processes and licensed manufacturers in many developed regions of the world. In parallel with this effort, a number of institutions independently developed variants of the technology, the most extensive and best known being operations within the former Soviet Union.

During those development years, EXW solutions were developed for a broad range of welding situations. These included micro-sized spot welds for the electronics industry, pipe-to-pipe butt welds for the gas transmission industry, tube-to-tube sheet welds for the power industry, simultaneously formed and cladded pots and pans, and several kinds of spot welds and overlapping butt welds. Although the EXW technology proved highly versatile technically, most of the applications did not prove to be commercially viable. The primary exceptions were clad plate manufacture and a variant process to make welding transition joints.

In today's industries, around 80% of the world's EXW production is clad plates, primarily used in the process industries for corrosion or wear-resistant equipment. The balance is mainly bimetal transition joints, used as junctions for making commercial fusion welds between 'non-weldable' metal components, predominantly aluminum-to-steel.

8.4 The EXW process

8.4.1 Process description

Figure 8.1 presents a schematic description of the EXW process which uses high energy from explosive detonation to produce a metallurgical weld between metal plates. The basic sequence of process operations is as follows:

1. Surfaces of metals are ground and the metals are fixed parallel with a predefined separation distance.
2. Especially formulated explosive powder is placed on the cladder surface.
3. Detonation front travels uniformly across the cladder surface from the initiator.
4. Cladding metal collides with backer at a specific velocity and impact angle.

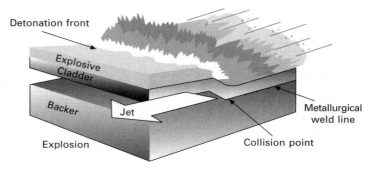

8.1 Schematic portrayal of the EXW process.

5   Momentum exchange causes a thin layer of the mating surfaces to be spalled away as a jet.
6   Jet carries spalled metal and oxides from the surfaces ahead of the collision point.
7   Thin layer of 'micro-fusion' 0.1 micron thick is formed at the characteristic wavy weld line.
8   Pressures exceeding 10,000 MPa hold the metals in intimate contact while metallurgical weld solidifies across the complete surface.
9   Rapid heating and cooling at the interface does not allow time for bulk heating of the metal. Total time above the melting point is in the range of 10 microseconds.
10 The EXW process assures that the backer materials retain specified physical properties and the cladding material retains the specified corrosion resistance properties.

8.4.2 Process advances

The current understanding of the EXW technology was relatively well developed by 1980. Blazynski[1] presents an excellent compilation of the technology developments through 1983. Since that time, the primary EXW process advances have been in the areas of safety, facilitation and industrialization.

Safety

The initial EXW development and production was performed using classical explosives, such as amatol or dynamite, which are relatively easily initiated (Class A in US-DOT terminology). These explosives exhibited good detonation characteristics and easy detonation velocity control. They had the downside of being initiation sensitive and required extensive safety measures during manufacture, transit, storage and use. By 1980, manufacturers were beginning to develop ANFO (ammonium nitrate fuel oil) blasting agents for large area EXW work. ANFO is more difficult to initiate than are Class A explosives. In US-DOT regulations it is a blasting agent, not an explosive. ANFO offers far greater safety and reduced regulatory control. Further, ANFO is less costly. The detonation characteristics of ANFO necessitated process modifications. Today, most commercial explosion welding companies use ANFO as their primary production explosive. Although ANFO is less easily initiated, when detonated the energy release can be highly destructive. Safety remains a major issue.

Facilitation

Owing to the large amounts of explosive involved in production work, often exceeding 1000 kg, for cladding large plates, EXW must be performed in an isolated and controlled environment. Traditional options were to work out-of-doors in a very isolated location or to work in an underground facility. In recent years, some practitioners have constructed vacuum chambers for EXW production. Working under vacuum conditions offers several operational and technical benefits. Further EXW production can potentially be performed in typical industrial facilities. Capital and maintenance costs have caused this variant of EXW production to be limited.

Industrialization

The major advances in EXW over the past two decades have been in the area of industrial application development. The development of increasingly broader applications for EXW clad plate products and the development of products derived from clad plates has resulted in a growing EXW industry worldwide. These products and their areas of application are described in the following section.

8.5 EXW applications

8.5.1 Explosion clad industry

The dominant application of EXW today is in the manufacture of large, flat clad plates. Table 8.1 presents a list of typical metal types which are supplied as clad plates.

Explosion clad plates are readily formed into cylinders and heads and fabricated into process equipment. Figures 8.2 and 8.3 show clad plates and pressure vessels fabricated from clad plates.

EXW is one of the three cladding technologies being used for equipment manufacture today, the others being hot rollbonding and weld overlay.[8] Both

Table 8.1 Typical metals supplied as explosion clad plates, listed in order of decreasing commercial usage. Any of the cladding metals can be clad to any of the base metals

Cladding metals	Base metals
Titanium	Carbon steels
Stainless steels	Alloy steels
Nickel alloys	Stainless steels
Aluminum	Titanium
Zirconium	Copper
Tantalum	Nickel alloys
Copper alloys	Aluminum

8.2 Titanium-steel EXW clad plates being prepared for shipment, 2100 mm × 8000 mm × (125 mm steel + 8 mm titanium Gr 17).

8.3 Stainless steel clad refinery column. 4.6 m ID × 35.3 m long × (100 mm Cr–Mo steel + 3 mm type 321 stainless steel).

Developments in explosion welding technology 235

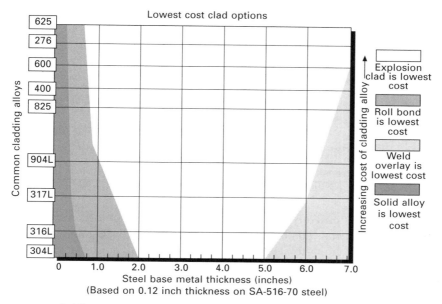

8.4 Comparative costs for clad products manufactured by EXW, hot rollbonding, and weld overlay as a function of thickness and cladding alloy cost (courtesy of Dynamic Materials Corporation).

rollbonding and overlay are limited to production of clad plates of compatible metal combinations, primarily stainless steels, nickel alloys and some copper alloys. Figure 8.4 presents the general competitive positions of the three technologies. EXW is generally the lowest cost technology when the cladding metal has a high cost and the base metal is in the mid-range (typically 50 to 100 mm) regarding thickness. When the cladding alloy and base are not metallurgically compatible, for example, titanium, zirconium, aluminum, or tantalum clad to steel, EXW is the only commercially significant technology for clad plate manufacture.

Explosion clad is used extensively in the manufacture of pressure vessels and heat exchangers for high pressure and/or high temperature corrosion-resistant processes. Applications are predominantly in the chemical, petrochemical, refinery, hydrometallurgy, and upstream oil and gas industries. For greater detail on these applications, equipment fabrication and clad performance issues, there are several good references.[9–13]

8.5.2 Welding transition joints

Welding transition joints, produced from EXW clad plates, provide a means for making fully welded dissimilar metal joints in normal fabrication environments. Figure 8.5 presents the transition joint concept. Transition

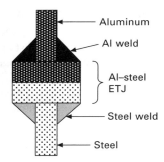

8.5 The transition joint (TJ) concept for using an EXW block to facilitate welding between aluminum and steel components.

joints are used extensively for joining aluminum to steel or to stainless steel. Other common metals combinations are aluminum-to-copper, aluminum-to-titanium, and titanium-to-steel or -to-stainless steel. Transition joints are primarily used as a replacement for mechanical connections in environments where mechanical joints have major technical weaknesses.

Primary applications include:

1 Ship construction for making high strength, crevice-free joints between aluminum bulkheads and steel decks. Corrosion maintenance costs are dramatically reduced.[14]
2 Truck and rail car installations for producing maintenance-free joints between light weight aluminum bodies and durable steel undercarriages.
3 Aluminum smelting plants for making resistance-free electrical connections between aluminum buss and steel anodes and cathodes.[15]
4 Other electrochemical systems for making maintenance-free electrical connections between aluminum and copper, aluminum and steel, or copper and steel.
5 Leak-free pipe couplings between aluminum and stainless steel, primarily for the cryogenic industries.
6 Leak-free pipe couplings between titanium and stainless steel for chemical process and aerospace applications.
7 Aerospace structural installations of high strength fully welded assemblies of Ti-6Al-4V joined to Inconel® 718.

8.5.3 Bipolar cell plates

Many compact electrochemical systems benefit from bimetallic, bipolar cell plates. Applications can range from chlorine and chlorate manufacture to speciality batteries and fuel cells. Titanium–steel and titanium–nickel bimetallic plates have been used extensively in chlorate and chlorine cells. EXW plates have generally been relatively thick and up to the present have found only limited uses in other bipolar cell equipment. In a variant of the EXW process, explosion clad plates are subsequently hot rolled to lighter gage plates. The technology, commonly referred to as 'bang and roll', is far more cost effective for manufacture of clad plate of total thickness less than 12 mm when large quantities are produced. The bang and roll technology has been used for production of large plates of titanium–steel, stainless and nickel alloys to steel, titanium–copper, and titanium–steel–copper.

8.5.4 Electronic packaging

Electronic component packaging often demands a number of unique performance requirements frequently not attainable in any single metal. Requirements can include electrical and thermal conductivity, thermal expansion, corrosion resistance and glass-sealing ability. EXW multilayer metal products offer an optimum technical solution in many situations. Examples include:

1. aluminum boxes with Kovar or stainless steel seal rings
2. copper boxes with Kovar or stainless steel seal rings
3. copper–Kovar or copper–molybdenum base plates
4. aluminum–Kovar glass sealing inserts
5. copper–Kovar bimetallic wire.

With the exception of the latter, all are manufactured by producing a relatively large clad plate, and then cutting it into many small, planar interface parts. Bimetallic wire has been manufactured by producing a fully bonded bimetallic cylinder and then by drawing to the required wire sizes.

8.5.5 Sputtering targets

The manufacture of modern electronic microprocessor circuits often includes sputtering processes. Typically the metal being sputtered, the target, is a high purity metal; examples include titanium, aluminum and tantalum. The targets are mounted in sophisticated machines, typically constructed of stainless steel, aluminum, and copper alloy components. Process efficiency necessitates that contact between the target plate and the base be electrically and thermally conductive, structurally sound and highly reliable. Explosion welding has been proved to be one of the most reliable target assembly processes. EXW manufacturing techniques are often specifically tailored to minimize consumption of the high cost target materials.

8.5.6 Thick plate manufacture

For some metal plate alloys, product thicknesses are limited by metallurgical processing considerations. For example, the relationship between centerline properties, plate thickness, and cooling rate considerations limits the thickness that can be produced in many aluminum and nickel alloys. Explosion welding of multiple plates together offers a way to produce a thicker product with reliable centerline properties. Production applications have included the production of a 300mm thick plate of 6XXX-series aluminum produced by EXW welding six 50mm plates together.

8.5.7 Partially welded plates: spot welds and line welds

Although EXW is typically used to produce large welded areas, the technology can be used for making line welds or spot welds. The explosive energy required is considerably lower. In the case of spot welds, techniques have been demonstrated for the manufacture of explosion welds under normal production shop conditions.[16]

8.5.8 Other applications

The applications presented above typify the broad range of products that have been produced using EXW and represent many commercially successful uses. Over the 40 years of its industrial development, EXW has been employed to obtain a much broader list of products. These range from a host of small, on-site applications to multilayer razor blades.

8.6 Weld characterization

The explosion welded interface typically exhibits a wavy morphology. When explosion welding paramenters are correctly selected, there is minimal evidence of melting along the back and crest of the wave. There is typically a swirl of material in the break of the wave which frequently contains solidified melt. Explosive energy and detonation rate have significant effect upon bond morphology. At low explosion detonation rates, the bond is typically flat. As the detonation rate is increased, the bond transitions from flat to wavy, and then at higher rates exhibits large waves and excessive swirling.[3] Explosion welding parameters and bond morphology have been shown to have a relationship which can be described in a manner similar to a Reynolds number. Bond zones will typically show no indication of melt layers when examined at the upper limits of present day optical microscopy. Further, there is no evidence of metal mixing or diffusion when examined at the current limits of commercial scanning electron microscopy. The bond appears to truly be a cold weld.

Transmission electron microscopy has been used to study the bond at a significantly higher magnification. At least three research groups have presented results from limited TEM studies of bond morphology.[17–19] All three studies

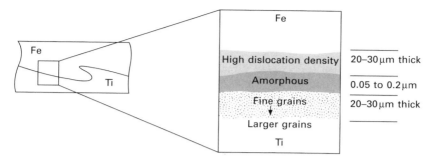

8.6 Schematic presentation of the high magnification appearance of a titanium–steel bond zone, indicating approximately 0.1 µm amorphous band at the EXW interface (Yamashita *et al.*[18])

found evidence of what appears to be prior molten metal in a region of 0.05 to 0.2 µm thick at the interface. This region exhibits mixing of the two metal types, but does not exhibit stable crystallographic or solidification structures. In the as-welded condition it appears to have a metastable amorphous atomic structure. The three studies have hypothesized that a thin layer of metal at the collision point has been heated well above the melting points and then resolidified at an extremely high cooling rate, in the range of 1×10^{-5} K/s. Under these conditions, there is insufficient time at a given temperature for steady state structures to form. If the interface is reheated to temperatures at which stable microstructures and intermetallics can form, the bondline gradually develops all of the thermally stable features that are observed in a steady state phase diagram between the component metals. Figure 8.6 presents a schematic of the as-welded bond zone region. These studies suggest the EXW weld is more realistically described as a hypercooled, micro-fusion weld. The cold weld characteristics of the interface result from the rapidity of the process, combined with highly localized melting and fusion.

8.7 Conclusions

EXW is a robust, well-developed welding technology. Its primary application is in the manufacture of clad plates and speciality products that are derived from clad plates, such as welding transition joints. The cold welding features of EXW provide unique capabilities for joining a large range of metals where traditional fusion welding technologies cannot be applied. Very high magnification analyses of explosion welds suggest that the unique interface conditions result from a hypercooled, microfusion weld. The cold welding characteristics of EXW are attributed to the rapidity of the heating and cooling rates, combined with highly localized melting and fusion.

8.8 References

1. Blazynski T.Z., *Explosive Welding, Forming, and Compaction*, Applied Scientific Publishers Ltd., Essex, UK, 1983
2. Holtzman A.H. and Cowan G.R., 'Bonding of metals with explosives', *Welding Research Council Bulletin*, 1965, **104**, April
3. Pocalyko A., 'Explosively clad metals' *Encyclopedia of Chemical Technology*, Vol 15, 3rd ed., John Wiley & Sons, 1981, 275–96
4. Banker J.G. and Reineke E.G., 'Explosion welding', *ASM Handbook*, Vol 6, *Welding, Brazing, and Soldering*, 1993, 303–5
5. Patterson A., 'Fundamentals of explosion welding', *ASM Handbook*, Vol. 6, *Welding, Brazing, and Soldering*, 1993, 160–4
6. Linse V., 'Procedure development and process considerations for explosion welding', *ASM Handbook*, Vol. 6, *Welding, Brazing and Soldering*, 1993, 896–900
7. Cowan G.R., Douglass J.J. and Holtzman A.H., US Patent 3 137 937, 'Explosive bonding', 1964
8. Smith L.M. and Celant M., *Practical Handbook of Cladding Technology*, Edmonton, Alberta, CASTI Publishing, 1998
9. Banker J.G., 'Try explosion clad steel for corrosion protection', *Chemical Engineering Progress*, AICHE, July 1996, 40–4
10. Banker J.G. and Winsky J.P., 'Titanium/steel explosion bonded clad for autoclaves and vessels,' *Proceedings of ALTA 1999 Autoclave Design and Operation Symposium*, Alta Metallurgical Services, Melbourne, Australia, May 1999
11. Banker J.G., 'Commercial applications of zirconium explosion clad', *Journal of Testing and Evaluation*, ASTM, W. Conshohocken, PA, March, 1996, **24**(2) 91–5
12. Banker J.G. and Cayard M.S., 'Evaluation of stainless steel explosion clad for high temperature, high pressure hydrogen service', *Proceedings of Hydrogen in Metals Conference*, Vienna, Austria. Oct. 1994, NACE International, Houston, TX
13. Banker J.G., 'Recent developments in reactive and refractory metal explosion clad technology', *NACE Paper 03459*, NACE International, Houston, TX 2003
14. McKinney C.R. and Banker J.G., 'Explosion bonded metals for marine structural applications', *Marine Technology*, Society of Naval Architects and Marine Engineers, July 1971, **8**(3), 285–92
15. Banker J.G. and Nobili A., 'Aluminum–steel electric transition joints, effects of temperature and time upon mechanical properties', in Schneider W. (ed.), *Light Metals 2002*, Warrendale, PA, The Minerals, Metals, and Materials Society, 2002, 439–45
16. Banker J.G., US Patent #6,772,934 Kinetic Energy Welding Process, 2003.
17. Chiba A., *et al.*, 'Microstructure of bonding interface in explosively-welded clads and bonding mechanism', *Materials Science Forum*, **465**, 465–74. Trans Tech Publications, Switzerland, 2004
18. Yamashita T., Onzawa T. and Ishii Y., 'Microstructure of explosively bonded metals as observed by transmission electron microscopy', *Transaction of Japan Welding Society*, Sept 1975, **4**(2), 51–6, Tokyo, Japan
19. Nobili A., Masri T. and Lafont M.C., 'Recent developmets in characterization of a titanium–steel explosion bond interface', Reactive Metals in Corrosive Applications Conference Proceedings, Haygosth J. and Tosdale J. (eds), Wah Chang, Albany OR, 1999, 89–98

9
Ultrasonic metal welding

K. GRAFF, Edison Welding Institute, USA

9.1 Introduction

The application of ultrasonic energy to materials joining processes has been in use for a number of years. While it was used for grain refinement in molten metals in the 1930s, for soldering, enhancement of resistance welding and in conjunction with arc welding in the 1940s, and for joining plastics in the 1950s, the ultrasonic metal welding process was first demonstrated in the early 1950s. It was found that ultrasonic vibrations were capable of creating a weld in metal parts without the need for melting the base metals.

The process of ultrasonic metal welding is one in which ultrasonic vibrations create a friction-like relative motion between two surfaces that are held together under pressure. The motion deforms, shears and flattens local surface asperities, dispersing interface oxides and contaminants, to bring metal-to-metal contact and bonding between the surfaces. The process takes place in the solid state, occurring without melting or fusion of the base metals.

Applications of ultrasonic welding are extensive, finding use in the electrical/electronic, automotive, aerospace, appliance and medical products industries, as examples. Although nearly all metals can be welded with ultrasonics, the widest current uses typically involve various alloys of copper, aluminum, magnesium and related softer metal alloys, including gold and silver. A number of dissimilar metal combinations (e.g. copper and nickel) are readily welded with ultrasonics. It is used to produce joints between metal plates, sheets, foils, wires, ribbons and opposing flat surfaces. Joining several stranded copper cables into a single junction, used in automotive wire 'harnesses' is one common use. Ultrasonic welding is finding increasing applications for structural components in the automotive and aerospace industries. It is very useful for encapsulating temperature sensitive chemical or pyrotechnic materials. The closely related process of ultrasonic microbonding is widely used in the semiconductor and microelectronics industries.

There are several variations of the welding process. The most common, ultrasonic spot welding, uses two different configurations of equipment, known as lateral drive and wedge-reed welding systems. Ultrasonic seam welding, used to obtain continuous welds in thin gage materials, and ultrasonic torsion welding, used for circular closure welds and stud welds, are added variations of the ultrasonic process, as is the previously noted ultrasonic microbonding. A number of other laboratory or experimental configurations of ultrasonic welding systems have been developed, including a means of producing ultrasonic butt welds.

This overview of the ultrasonic metal welding process will:

- Outline the principles of ultrasonic metal welding;
- Describe the features of several types of ultrasonic welding equipment;
- Summarize information on the mechanics and metallurgy of the ultrasonic weld;
- Review a number of applications of the process;
- Summarize advantages and disadvantages of the process;
- Examine future trends;
- Provide references for further study of the process.

9.2 Principles of ultrasonic metal welding

It was noted that ultrasonic welding uses high frequency mechanical vibrations to create a friction-like relative motion between two surfaces, causing deformation and shearing of surface asperities that disperse oxides and contaminants. This process brings metal-to-metal contact and bonding between the surfaces. Its features may be illustrated by considering a typical welder, known as a lateral drive system, as shown in Fig. 9.1.

The key elements of the system are the transducer, booster and sonotrode series of components, which produce and transmit the ultrasonic vibrations to the workpieces clamped between the sonotrode welding tip and a rigid anvil. The static force between the anvil and sonotrode tip is produced by a coupling moment created within the system enclosure by an arrangement of leveraged forces and pivots (details not shown in this simplified schematic).

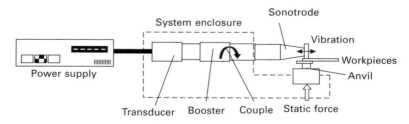

9.1 Lateral drive ultrasonic welding system.

The high frequency (e.g. 20 kHz) vibrations are produced in the transducer via a high frequency input voltage from the electronic power supply.

Since ultrasonic vibrations are the basis of this welding process, a basic understanding of the underlying vibrational behavior of the transducer–booster–sonotrode-system is important. This requires starting with the most important component, the ultrasonic transducer, shown in Fig. 9.2. Thus, from Fig. 9.2, we see that:

- The main components of the transducer are a front driver (usually aluminum or titanium), several piezoelectric disks (in pairs of two, four or six – sometimes up to eight), and a rear driver (usually steel), with this assembly held together by a bolt that serves to precompress the ceramics. The piezoelectric properties of the disks result in conversion of the high frequency electrical signal from the power supply to mechanical vibrations of the transducer. Between the disks are thin steel, copper or nickel electrodes which connect the disks to the external power supply.
- The transducer assembly vibrates, in a longitudinal direction (i.e. along the axis of the transducer, as shown by the arrow) at a resonant frequency determined by the dimensions of the drivers and ceramics. The distribution of vibration is shown by the curve labeled 'vibration half wavelength', indicating that the maximum amplitude is at the front end of the transducer. There is a point of minimum vibration, known as the 'node' which is usually designed to be located at the flange, where the transducer enclosure case may be attached, as shown in Fig. 9.2.
- The dimensions and materials of the various transducer components are selected so that the device vibrates, or more specifically 'resonates' at a specific frequency. An operating frequency of 20 kHz is commonly used in many systems, but transducer frequencies as low as 15 kHz or as high as 60 kHz may be found in metal welding systems (and yet higher, e.g. 120 kHz to 300 kHz, in electronic microbonding systems). An important

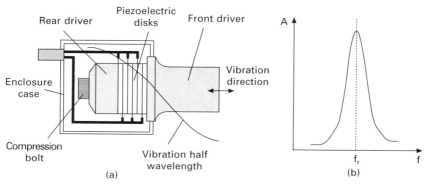

9.2 Ultrasonic transducer: (a) transducer assembly; (b) resonance behavior.

characteristic of the transducer is that its operating frequency (e.g. 20 kHz) is very sharply tuned to that specific frequency, with its vibration amplitude dropping off extremely rapidly just a few hundred hertz from its operating point. This is shown in Fig. 9.2(b), where the amplitude of transducer vibration is plotted against frequency and shows how just a slight shift from its resonant frequency, f_r, results in a dramatic reduction in amplitude.

- The amplitude of vibration of the front end of the transducer is quite small, typically in the range of 10–30 µm, peak-to-peak, an invisible amount to the unaided eye. This, combined with the fact that operating frequencies are typically above the audible limit, results in little visible or audible action during a welding cycle, thus being deprived of the spectacular pyrotechnics of arc, resistance and laser processes.

The three components of transducer, booster and sonotrode, are shown assembled in Fig. 9.3 (the assembly is by threaded fasteners at the component interfaces). While each differs in shape from the transducer, the booster and sonotrode are in longitudinal vibration, tuned to the same frequency as the transducer, and with a node in the middle region of each. Each is operating at a half acoustical wavelength with the result that the overall assembly is $3 \times {}^1\!/_2 = 1 + {}^1\!/_2$ (1.5) wavelengths long. The exact value of the acoustic wavelength depends on the frequency, the materials of the different parts and their geometric shapes. To an approximation, at 20 kHz, the acoustic wavelength in several materials used in practical systems (e.g. steel, aluminum, titanium) is 25 cm, so that the length of the system shown in Fig. 9.3 would be of the order of $1.5 \times 25\,\text{cm} \approx 38\,\text{cm}$. The booster and sonotrode are shaped to amplify the vibrations of the transducer, so that as one progresses down the length of the assembly, the amplitude is first increased by the booster and then again by the sonotrode. The result is that an amplitude at the transducer, possibly of 20 µm, may be increased to as high as 100 µm at the weld tip on the sonotrode. The welding tip of the sonotrode may be detachable, held in place in a threaded connection, as suggested by Fig. 9.3, or be machined as an integral part of a single piece sonotrode.

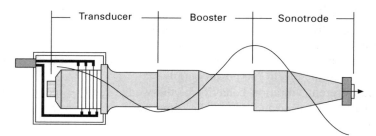

9.3 Transducer, booster and sonotrode of the lateral drive welding system.

The electronic power supply, shown in Fig. 9.1, provides the driving power to the system, converting line frequency to the ultrasonic frequency required by the transducer. Depending on the power rating of the transducer, and the application, the supply may need to provide power levels of tens to thousands of watts. The power supply system also provides additional control and operation functions. The piezoelectric transducers typically have a high level reactive (of a capacitance nature) electrical impedance of up to several hundred ohms which would match poorly with modern electronic power supplies that are designed for low impedance (e.g. 50 Ω) resistive loads. For this reason, impedance matching circuitry is typically built into the power supply.

Of equal importance is the fact that the sharply tuned resonant frequency of the transducer–booster–sonotrode system will change during system operation, and from one welding application to another. As one example, continued operation of the system will cause heating of the various parts, resulting in a change in system frequency. The welding application serves to impose a mechanical impedance on the transducer system through the welding tip, with this also being capable of changing the system resonant frequency. These and other effects can result in shifts in the sharp resonance curve shown in Fig. 9.2(b) away from the drive frequency and could greatly reduce vibration amplitude. However, all modern welding power supplies use frequency tracking circuitry to compensate for such shifts, so that the power supply frequency will stay tuned on the changing system resonant frequency. In addition to this, some welding power supplies are able to control the vibration amplitude of the transducer during the weld cycle, with this typically being done using concepts based on equivalent circuit representation of the transducer.

A second type of widely used configuration for ultrasonic metal welding is known as a wedge-reed system. Although using different principles to impart ultrasonic vibrations to the workpieces, it ends up achieving the same effect of a frictional-like relative motion at the surfaces of the parts. A wedge-reed system is shown in Fig. 9.4(a). The key elements of this system are the transducer, wedge and reed series of components, which produce and transmit the ultrasonic vibrations to the workpieces clamped between the sonotrode welding tip and an anvil. A static force is applied between the mass at the upper end of the reed and the anvil. As with the lateral drive, the high frequency transducer vibrations are produced by a high frequency input voltage from the electronic power supply.

The different vibration behavior of the wedge-reed system is shown in Fig. 9.4(b). The transducer is of the same form as that in the lateral drive system. The wedge serves the same purpose as the booster in the lateral drive system, acting to increase transducer vibration amplitude. Thus, the transducer-wedge is in longitudinal resonant vibration, as shown by the dashed wavelength

246   New developments in advanced welding

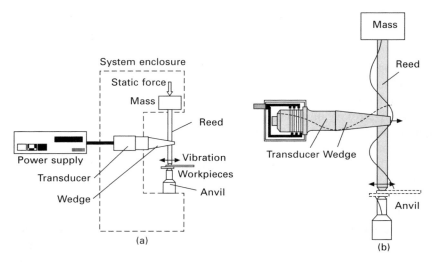

9.4 Wedge-reed ultrasonic welder: (a) overall system; (b) transducer, wedge, reed and anvil.

pattern in the figure, and produces vibrations in the direction of the arrow at the tip of the wedge. The wedge is solidly attached to the vertical reed by welding or brazing. The different feature of this system is that the wedge vibration drives the reed in a bending, or flexural, vibration mode. The general nature of the vibration pattern along the reed is shown by the solid line wave pattern. Being in bending vibrations means that the motion of the reed at any point is in a left–right (or transverse) direction, much like the vibrations of a string. This results in a transverse vibration of the welding tip against the workpieces, as shown in Fig. 9.4. Thus, the wedge-reed produces the same vibration effect at the workpieces as in the lateral drive (see Fig. 9.1), but by a slightly different means. Another variation of the lateral drive system involves the anvil, which also is in vibratory bending motion, as shown in the figure, although in some cases, a rigid anvil design may be used. Using a 'contra-resonant' design, the anvil may vibrate out of phase with the reed, thus increasing the net transverse motion across the parts.

With this outline of the vibration principles that underlie ultrasonic welding, we now can examine more closely the weld itself, shown in Fig. 9.5(a). It has been noted that both of the preceding welding systems end up applying the same type of transverse vibration from the weld tip to the top surface of the workpiece, as shown in Fig. 9.5. The top part moves in unison with the welding tip. This 'anchoring' of the part to the tip is usually assisted by a roughened or knurled surface to the tip, to engage better with the surface. Similarly, the bottom part remains anchored to the anvil, also usually assisted by a patterned anvil surface. As a result, the motion of the welding tip produces a relative motion between the parts at the contact surfaces. This

Ultrasonic metal welding 247

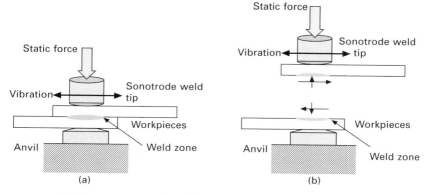

9.5 Ultrasonic metal weld: (a) transverse vibration imparted to workpieces, and weld zone; (b) normal and shear forces acting at weld zone.

relative, transverse motion, between the two opposing surfaces creates the friction-like action. This action, in turn, causes shearing and plastic deformation between asperities of the opposing surfaces, bringing about increasing areas of metal-to-metal contact and solid state bonding between the parts. The actual bonding zone is suggested by the shaded region between the two parts. If the two parts are separated at the faying surfaces, as shown in Fig. 9.5(b), two types of force components will be seen acting at the interface. First is the static clamping force, acting at right angles to the surfaces, and the second is a transverse shearing force, brought about by the friction-like transverse motion of the parts. These forces will be explored in more detail in Section 9.4.

Two unique features of ultrasonic metal welding are worth re-emphasising:

1. The nature of the motion between the parts during the metal welding process is one of transverse oscillation, where the opposing surfaces move parallel to one another. This is in distinct contrast to the allied process of ultrasonic plastic welding, where the opposing surfaces move at right angles to one another.
2. Although there is a local plastically deformed weld zone created between the parts, no melting of the metals occurs in this zone. The bonding is via a solid state bond, versus a fusion bond such as occurs in the weld nugget of a resistance spot weld, or in other fusion-based processes such as arc or laser welding.

Additional features of the overall bonding mechanism of the ultrasonic metal weld will be discussed in Section 9.4.

From the illustrations of the lateral drive and wedge-reed welding systems and of the weld zone it is evident that a number of parameters can affect the welding process. The main ones can be summarized as follows:

- ultrasonic frequency
- vibration amplitude
- static force
- power
- energy
- time
- materials being welded
- part geometry
- tooling.

Not all of these are independent parameters. For example, the energy delivered to the welder is dependent on the power–time relationship, while the vibration amplitude may depend on the power level. These are all noted here separately because different welding systems may be set up to operate with differing sets of independent variables.

9.2.1 Ultrasonic frequency

It has been noted that ultrasonic welding transducers are designed and tuned to operate at a specific frequency, with these frequencies ranging, for different systems and applications, from 15 kHz to 300 kHz. Most metal welding systems will operate at 20, 30 or 40 kHz, with the higher frequencies, e.g. 60 or 120 kHz, being used in microbonding work.

It is sometimes asked whether there might be 'critical frequencies' from a fundamental metallurgical physics point of view (e.g. exciting dislocation fields) at which the welding process might be optimal. There is no evidence of such material behavior, at least in the frequency regimes of ultrasonic welding. Instead, welding frequency is governed by such matters as welding power requirements, which are governed by part dimensions and materials being welded, and the overall design of transducers and coupling components.

While reference has been made to the 'single, specific operating frequency' of a transducer, and the nature of this sharply tuned resonant behavior illustrated in Fig. 9.2(b), in actual operation, there are several factors that act to shift the transducer resonant frequency. Small dimensional changes due to system heating during operation, varying static force, different tooling and changing tool conditions due to wear, and the changing effects of the welding load can all act to cause both long-term, as well as quite rapid, changes in the system resonant frequency.

However, while these shifting conditions of resonance can be complex, the key point from a practical user's standpoint is that modern power supplies employ sophisticated feedback control circuitry that automatically compensates for these shifting conditions and tracks the driving frequency of the transducer to maintain the system on resonance. Thus the changing conditions of welding effectively become transparent to the machine user, while the system is kept at resonance.

9.2.2 Vibration amplitude

The vibration amplitude of the welding tip at the weld is one of the key parameters affecting welding. Thus, it directly ties in to the energy delivered to the weld zone. It is again pointed out that ultrasonic vibration amplitudes are quite small quantities, being of the order of 10–50 µm at the weld and seldom exceeding 100 µm as a maximum. (The diameter of a human hair is in the range of 75–100 µm.)

In some welding systems, the amplitude is a dependent variable, being related to the power applied to the system. In other systems, the amplitude is an independent variable, capable of being set and controlled at the power supply because of added features of the feedback control system. (Interestingly, the purely mechanical parameter of vibration amplitude is controlled by purely electronic means, using certain equivalent circuit concepts between mechanical and electrical systems.) The selection of weld vibration amplitude will depend on the conditions of welding as governed by materials and tooling.

9.2.3 Static force

The static force is also a key parameter of ultrasonic welding. The force that is exerted on the workpieces via the welding tip and anvil, in pressing the parts firmly together, creates intimate contact between the opposing surfaces as preparation for the ultrasonic vibrations in the weld zone. The magnitude of the force will be strongly dependent on the materials and thicknesses being welded, as well as on the size of weld being produced and may range from tens to thousands of newtons. For example, producing a weld of 40 mm^2 in 6000 series aluminums may use forces of the order of 1500 N, while 10 mm^2 welds in 0.5 mm thick soft copper sheet may require only 400 N. In adjusting system parameters, there is typically found to be an optimum range of static force, below which welds will be weak to non-existent and above which excessive deformation of the parts may occur.

9.2.4 Power, energy, time

While individually listed as separate weld parameters, these are most conveniently examined in a unified manner, since they are all tied closely together. When a weld is made, the voltage and current result in a time-varying electric power flow to the transducer over a period of the weld cycle. This power–time curve can take many forms, depending on material types,

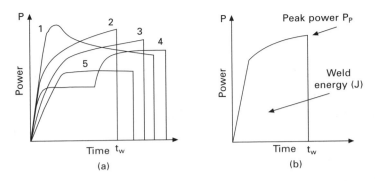

9.6 Weld power: (a) examples of weld electrical power curves; (b) representative power curve.

dimensions and surface finishes, ultrasonic parameters such as amplitude and static force, and the particular features of a given welding system. Figure 9.6(a) is simply representative of the various forms that power curves can take during welding. However, these many details may be ignored for present purposes, and the simple curve of Fig. 9.6(b) used to illustrate the basic power, energy and time relations that would apply to more complex shapes.

A representative power curve, Fig. 9.6(b), will have a peak power (P_P) and weld time (t_w). The area under the power curve is the weld energy (joules) or, more specifically, the electrical energy supplied to the transducer during the weld cycle. It is evident that not all three can be independent. Thus, in Fig. 9.6(b) one can set the peak power, with the welder running until that level is reached, with weld energy and time being dependent on reaching a certain level. Or, energy can be set and the weld would run until such as the set level is achieved, and so forth for other variations.

The power delivered to the transducer from the power supply is converted to ultrasonic power at the weld. However, between the electrical input and the weld several conversions and transmission steps intervene. The actual power delivered to the weld zone is dependent on several factors that can include: (a) the efficiency of electromechanical conversion of the electrical input to mechanical output by the piezoelectric materials; (b) losses in the bulk materials and at interfaces of the transducer–booster–sonotrode system; (c) power radiated from the weld into the workpieces and the anvil structure. The power setting may be indicated in terms of the high-frequency power input to the transducer, or the load power (the power dissipated by the transducer–sonotrode–workpiece assembly). Currently, estimates of actual power (and energy) to the weld itself are made by measuring input electrical power and subtracting off estimated system losses determined from no-load system measurements.

The weld time of ultrasonic metal welds, as noted, may be a dependent or independent variable based on the type of welding system being used. By whatever means, however, metal welding times are quite short, typically being well under 1 s in duration – thus 0.25 to 0.5 s are common. Longer welding times usually suggest the need to examine and possibly modify system parameters.

9.2.5 Materials

The single category 'materials' in fact encompasses a wide range of issues and parameters relating to ultrasonic metal welding. First, of course, is the type of material. As will be noted in Section 9.4, claims have been made that nearly every metal can be ultrasonically welded in some fashion. The properties of the materials, including modulus, yield strength and hardness are a key consideration. Generally speaking, the softer alloy metals, such as aluminum, copper, nickel, magnesium and gold/silver/platinum are most easily welded ultrasonically. With increasing alloy hardness, ultrasonic welding increases in difficulty. The material surface characteristics come next, with these including finish, oxides, coatings and contaminants. Given that the ultrasonic welding process by its very nature involves the transverse vibration of opposing surfaces, held together under pressure, it is evident that the conditions of these surfaces will play an important role.

9.2.6 Part geometry

The shape of the parts to be welded is also important. The dominant feature here is that of part thicknesses. Simply put, the thinner the parts, of whatever material, the better the chances of achieving ultrasonic welds. Increasing part thickness, in particular that of the part in contact with the welding tip, requires larger welding tip areas, increased levels of static force and generally increasing weld powers. Maximum thicknesses that can be achieved will obviously depend on the material being welded and the power levels available from a welder. For example, welding 1–2 mm 5XXX, 6XXX aluminum alloys is achievable with welding systems in the 2.5–3.5 kW ranges. Another part geometry factor, becoming increasingly appreciated in welding larger parts, is the overall lateral size (width, breadth) and shape of the parts. The very vibrations that create an ultrasonic weld can, when transmitted away from the weld and into the surrounding part, affect the making of the weld itself, as well as affect previously made welds. Issues of part resonance, which may play little role for small parts, can become a factor in larger parts where dimensions may become of the order of ultrasonic vibration wavelengths. Typically, issues in this area can be accounted for by modifying part dimensions and stiffness, as well as use of supplemental clamping points to dampen vibrations.

9.2.7 Tooling

The tooling consists of a sonotrode/welding tip that contacts the top surface of the top part, and an anvil that contacts the bottom surface of the bottom part. The tooling serves as part support and to transmit ultrasonic energy and the static force to the parts being welded. In most cases, detachable tool tips and replaceable anvils are used on ultrasonic welders. In some cases, the tool tip is machined as an integral part of a solid sonotrode. While the weld tip and anvil contact surfaces are usually flat, in some cases the weld tip may be designed with a slight convex curvature in order to change the contact stress patterns. The tooling contact surfaces typically have machined knurled patterns of grooves and lands, or other surface roughening, to improve gripping of the workpieces. A wide range of tooling hardnesses, heat treatments and coatings are employed to deal both with the wide range of materials to be welded and the wear conditions that are encountered for varied applications.

9.3 Ultrasonic welding equipment

In reviewing the principles of ultrasonic metal welding, the basic features of two of the most widely used systems, the lateral drive and the wedge-reed, have been described and shown in Fig. 9.1 and 9.4. An example of a 20 kHz, 2.5 kW wedge-reed welder is shown in Fig. 9.7. These various systems produce spot welds whose areas depend on the specifics of materials, thicknesses and welder power capabilities. Thus, 2.5 kW–3.0 kW, 20 kHz welders can produce spot sizes of the order of 40 mm^2 or larger.

Another type of weld that can be achieved with ultrasonics is a seam weld. The basic features of an ultrasonic seam welder are shown in Fig. 9.8, where a continuously turning, lateral drive type of ultrasonic transducer vibrates a circular disk sonotrode that traverses the workpieces, producing a continuous weld. The details of fixturing and moving the rotating transducer system, including bearings and drivers, are not shown in this simplified diagram. The disk is machined as an integral part of the sonotrode, which requires turning the entire transducer–sonotrode assembly. A fixed anvil is shown in the figure, but in practice the anvil may be moving in unison with the turning sonotrode disk. The means of achieving this synchronous motion may be in the form of a turning disk anvil, or simply a laterally moving flat anvil base. In some cases, the rotating ultrasonic transducer is fixed in space, or it may be traversed along the workpiece. As with spot welders, the transducer must be leveraged to apply a static force to the workpieces as it turns. Seam welders find extensive use in joining aluminum and copper foils.

Another means of achieving an ultrasonic weld is to impart a twisting or torsional motion to specially designed horns and tooling, thus producing an ultrasonic torsion weld (sometimes called a 'ring weld'). The means of

Ultrasonic metal welding 253

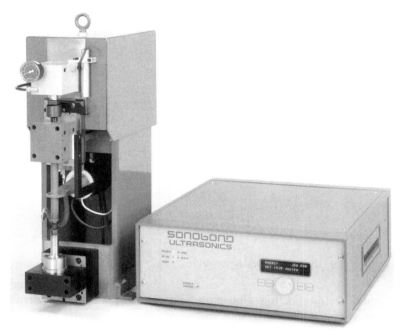

9.7 Wedge-reed ultrasonic welding system (source: Sonobond).

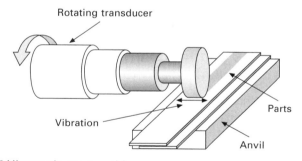

9.8 Ultrasonic seam welder.

achieving this is shown in Fig. 9.9, where two ultrasonic transducers, operating in longitudinal vibrations as previously shown in Fig. 9.2, are attached (typically they are welded) to ultrasonic boosters and tooling to produce a push–pull or torsional vibration to the booster system. The booster–tooling is specifically designed to be resonant in the torsional mode, versus the longitudinal mode, as used in the previous systems. The torsional resonance produces a circular vibration at the sonotrode and welding tip, which in turn creates a circular weld pattern on the workpiece. Although the vibration is of a circular nature, the motion of the tool surface is still parallel to the workpiece surface, as in

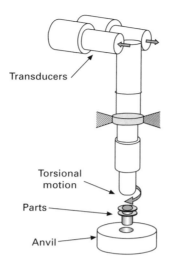

9.9 Ultrasonic torsion welder.

the spot welding action. Hence, the nature of the weld that is created is the same as that produced in spot and seam welding.

The concept of the torsion welder has been illustrated for two transducers. Depending on power requirements, from one to four transducers may be used. A four transducer torsion welder can produce up to 10 kW of power at 20 kHz.

9.4 Mechanics and metallurgy of the ultrasonic weld

In examining the mechanism of ultrasonic metal welding, it is helpful to start with the weld zone, as first depicted in Fig. 9.5(a), where the workpieces are shown adjacent to each other and Fig. 9.5(b) where they are shown in separation. This separation along the plane of the pre-welding interface also displays the primary forces present in making an ultrasonic weld, namely the shearing force, caused by the transverse ultrasonic vibration of the parts, and the normal force, caused by application of the static clamping force. The shear, normal force vectors are, of course, the result of a distribution of normal and shear stresses over the contacting surfaces of the two parts.

Now consider the condition of the two surfaces that will be in the zone of welding. Examined on a magnified scale, it is realized that the opposing surfaces consist of peaks and valleys whose profile depends on the surface finish of the materials. It is further realized that the surfaces when pressed in contact, will initially only be in contact at intermittent asperities or 'high spots.' While the number of contact points will depend on surface roughness

Ultrasonic metal welding 255

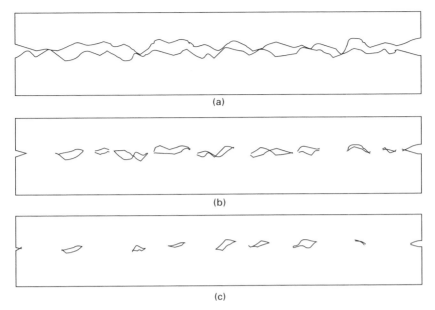

9.10 Development of contact surfaces in ultrasonic welding: (a) initial contact through asperities; (b) and (c) progression of shearing, deformation and formation of weld zones as the weld develops.

and clamping force, the general nature of initial static contact between a small region of the surfaces within the weld zone is shown in Fig. 9.10(a). Further, it is realized that the surfaces have oxide coatings, as well as possible surface contaminants, such as finishing or forming lubricants and absorbed moisture, which generally prevents a pure metal-to-metal contact between the surfaces, although at some contact points there may be penetration of oxides and local microwelds might occur.

When the ultrasonic vibrations are started, the top piece moves relative to the bottom piece in a transverse, friction-like motion. Plastic deformation and shearing of the interfering asperities occurs, cutting through surface contaminants and fracturing and dispersing surface oxides, resulting in increased metal-to-metal contact across the surfaces and formation of weld zones (also called 'microwelds'). Continued vibrations result in increased areas of contact, until complete or nearly complete contact and joining of the surfaces has occurred and a weld between the parts developed.

Numerous studies of the progression of this weld process have been done by stopping the weld cycle at various stages and peeling apart the surfaces. These studies show initial intermittent 'islands' of bonded surface. Initial bonding may also occur around the circumference of the weld. This particular feature is a consequence of using a welding tip that has a shallow spherical curvature, which creates a contact stress distribution that has a maximum at the circumference. Flat weld tips do not exhibit this feature. The short, elongated striations visible at the start of the process correlate with the direction of ultrasonic vibration, but not with the amplitude, which may only be on the order of 10–15 µm at the interface. They instead represent the growth of microwelds from initial contacts. Progressive growth of the microwelds occurs until a point is reached at which a nearly complete weld is achieved.

The nature of the bond that is formed across the interface of the parts is solid state; that is, it has been achieved without melting and fusion of the workpieces, but instead brought about by direct metal-to-metal adhesion of the solid materials. While the bond is solid state, this does not suggest that temperature does not play a role in the process. The plastic deformation that occurs results in a noticeable temperature rise, with this rise varying with materials and welding conditions, but always being below melting temperatures. However, the yield point of materials is temperature sensitive, and it is found that ultrasonic welding temperature increases are sufficiently great to cause reduction in the local yield strength of materials in the weld zone. This reduction in turn enhances further plastic deformation and flow of the materials in the weld zone.

It should be noted that a number of studies have been done on the temperatures of the weld zone and at the weld interface, typically using thermocouple and infrared techniques. Temperatures were generally found to rise very quickly in the initial welding stage, then to remain stable for the remainder of the cycle. Temperature rises varied greatly with the metals and metal combinations being welded. The heat generated by plastic deformation is quite localized at the interface and may be sufficient to cause recrystallization and diffusion. Studies of aluminum welds, for example, showed maximum temperatures reached to be on the order of 400 °C. In general, the temperatures reached during welding will depend on the mechanical properties of the harder/stronger of the materials welded. Thus, temperatures of a copper–Monel weld would be higher than those of a simple copper–copper weld.

This general description of the weld process has been basically a mechanistic one, where local vibration, plastic deformation and heating create the circumstances for a solid-state metallurgical bond to be achieved. Metallurgical examination of the weld zone shows local plastic deformation to be confined to a very small thickness, as brought out by Fig. 9.11 for the case of a weld in 6061T6 aluminum which shows a typical section of a completed ultrasonic weld. The 'sawtoothed' upper surface is a result of the weld tool imprint on the top weld part. The thickness of the actual weld zone of deformed material is quite small, approximately 50 µm for this particular case. This layer consists of very fine grain structure resulting from heavy plastic deformation. After 0.1 s of weld time (i.e. approximately a third of a 0.3 s weld cycle), one can

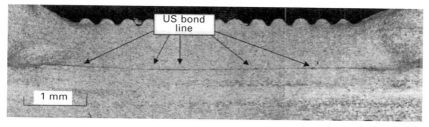

9.11 Ultrasonic weld metallography: entire interface (source: de Vries, 2000).

see discrete deformation islands or microwelds separated by unbonded surface. At completion of a 0.3 s weld, one arrives at a continuous interface layer. The grain structure just a short distance from the weld is substantially undisturbed, while an irregular thin zone of fine or even amorphous structure exists in the zone. There is some undulation of the pattern, which is sometimes seen to transfer to near-turbulent patterns of mixing in some regions of the zone.

Given this background information on the mechanism of ultrasonic welding, and some examples of metallurgical features, it is natural to inquire as to what metals are ultrasonically weldable. 'Weldability' as defined by The American Welding Society refers to the ease with which materials may be joined to meet the conditions of their intended service. Over the years, a large number of materials and material combinations have been investigated for their ultrasonic weldability. As a starting guideline, the chart shown in Fig. 9.12 may be used. Thus, metals that have been shown to be weldable are listed along the top of the chart and range from aluminum to zirconium. These same materials, or their alloys, are shown along the chart diagonal. Most materials can be joined to themselves or their alloys, that is, 'monometal' welds. Exceptions to this are germanium and silicon. It is seen that aluminum is exceptionally weldable, having been joined to all the listed metals. Other easily welded materials include copper alloys and the precious metals (gold, silver, platinum). On the other hand, iron and its alloys including steels, and refractory metals, such as molybdenum and tungsten, while weldable, can typically only be used in thin gages. Welding of softer alloys to these materials is more readily accomplished, however. In general, it is found that ease of weldability is tied to ease of plastic deformation, so that material hardness and yield strength play important roles in ultrasonic welding, with higher strength and hardness materials being increasingly difficult to weld.

While attention here has been on metal weldability, it should be noted that ultrasonic welding has been used to achieve joints between metals and non-metals. In particular, metals have been joined to ceramics and types of glass. Typically, the metals are in the form of foils and are those of more easily bonded metals (e.g. aluminum, copper). In most cases a thin foil transition layer is used, or a thin metallization has been applied to the glass or ceramic substrate.

258   New developments in advanced welding

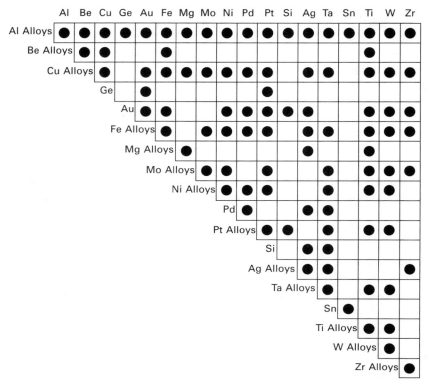

9.12 Ultrasonic material weldability (source: American Welding Society, 1991).

In addition to the limits imposed by materials on the scope of ultrasonic welding, there are limits imposed by the geometry and dimensions of the parts being welded, chief of which are thickness limitations. Thickness limits can be understood in terms of the contact stresses acting between the surfaces being welded and their relationship to weld tip geometry and thickness of the top welded part.

Using Fig. 9.5(b) as a starting point, the general case of a flat welding tip is illustrated in Fig. 9.13(a), where interface shear stresses and the vibration component of the weld tip have been omitted for simplification. Thus, the contact stresses between the flat weld tip and top surface are shown as approximately uniform. These contact stresses are transmitted to the weld interface, spreading out somewhat to cover a wider contact area, with this resulting in a reduced contact stress amplitude and a smoothing out of the edges of the stress distribution.

If the top part is very thin, there will be very little reduction of contact stress between the tip and the interface, as brought out in Fig. 9.13(b–1),

Ultrasonic metal welding 259

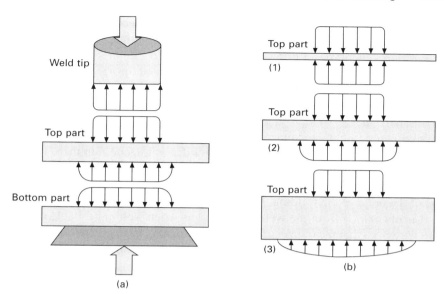

9.13 Contact stresses in weld zone: (a) stresses at weld tip, top and bottom parts; (b) stresses on top part with increasing part thickness (1)–(3).

where just top part stresses are shown. On the other hand, if the top part is quite thick (with 'thin' and 'thick' being measured relative to the width of the weld tool), then the amplitude and shape of the contact stresses may be greatly changed. These situations are shown in Fig. 9.13(b), where the focus is reduced to just the top part. Thus, in the thin top part of Fig. 9.13(b–1), the contact stresses are little changed in transmission to the interface. Figure 9.13(b–2) is the situation shown in Fig. 9.13(a). The case of a thick top part is shown in Fig. 9.13(b–3), where the form and magnitude of the contact stresses are greatly changed at the interface.

Now, under ultrasonic vibration, the contact stresses directly influence the weld producing interface shearing stresses. If, for a given thickness, the total static clamping force is too small, the resulting contact stresses, and resultant shearing stresses, may be insufficient to create a weld. Obviously an increase of static force will increase contact stresses, but if part thickness is too great, the amount of deformation at the top surface may be excessive, before conditions of welding are reached at the interface. The general relationship of part thickness, weld tip size and contact stresses is important in understanding the basic factors influencing welding. Thus, increased part thickness requires an increased size of weld tip in order to maintain uniformity and level of stresses at the interface. Increased weld tip size in turn demands increased clamping forces to maintain stress levels for welding. These in turn demand greater power of the ultrasonic welding system in order to drive the welding tip, under high forces, at the vibration amplitudes needed to achieve the necessary shearing stresses and welding action at the interface. While this provides a general description of the issues of part thickness and welding, there are no current formulas that relate these issues of tip size, part thickness and materials to weldability.

Another issue of welding relates to overall lateral dimensions and shape of the parts being welded. It is evident that achieving an ultrasonic weld relies on vibrating the top piece relative to the bottom piece. Some motion is absorbed in the local plastic deformation at the weld zone, but there is also some net motion imparted to the top part. In many applications of ultrasonic welding, the dimensions and mass of the top part are small, i.e. dimensions are of the order of the welding tip (e.g. electrical contacts), or the mass of the part is uncoupled from the weld zone (e.g. as when welding flexible cables and wires), so that the impact of top part motion may be neglected. However, in applications where top part dimensions begin approaching the longitudinal vibrational wavelength in that material, the forces acting in the weld zone may be affected by top part mass and dimensions. In particular, the interface shearing forces may be impacted by these vibration phenomena, dropping dramatically and ceasing to create a weld. In general, when welding larger parts that have dimensions of the order of acoustic wavelengths, attention must be paid to part vibrations as they may affect the weld being made or welds previously made.

9.5 Applications of ultrasonic welding

The applications of ultrasonic metal welding are widespread, but have greatest use in all aspects of electrical and electronic connections as found in several industries including electrical, automotive, medical, aerospace and consumer goods.

Thus, in the automotive and trucking industries, wire harnesses serve to distribute electrical signals and power to locations throughout the body structure, and require multiple branches and consolidations of braided and solid wires of various sizes. Similar applications, although on a smaller scale, are found in the appliance industry. Figure 9.14 is an example of a typical consolidation of several leads to a common junction, while Fig. 9.15 shows various multiple wire junctions common in wire harnesses. Sensor terminations, contact assemblies, braided wire connections and buss bar terminations are all examples of connections found in the automotive and trucking industries. Electric motors, field coils, transformers and capacitors are other examples where ultrasonic welding is used in their assembly. The use of ultrasonic micobonding for microelectronic interconnections remains one of the most extensive uses of ultrasonic welding.

Battery and fuel cell manufacture uses ultrasonics to make various joints

Ultrasonic metal welding 261

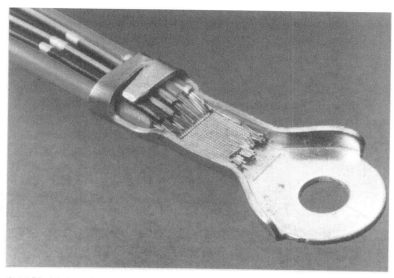

9.14 Multiple wire to single terminal (source: Telsonic).

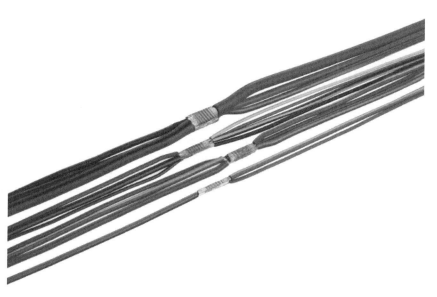

9.15 Various wire harness multiple wire junctions (source: Telsonic).

in these products, involving thin gauge copper, nickel or aluminum tabs, foil layers, or metal meshes and foams. Both spot and seam welding are widely used.

Packaging applications are another field where ultrasonic welding is applied widely using seam, torsion or conventional spot welding systems. For example,

262 New developments in advanced welding

Fig. 9.16 Ultrasonically welded aircraft access panel (source: Sonobond).

seam welding is widely used to seam foil food and cooking pouches hermetically and also finds application for splicing foil rolls during their manufacture. Torsion welding is used to seal a wide variety of cylindrical containers, where the contents may be highly reactive or heat sensitive (e.g. air bag igniters), as well as making stud weld attachments. Other packaging uses include sealing of tubes in the refrigerant and air conditioner industries.

A future trend in the use of ultrasonic welding will be in structural automotive and aerospace applications, joining thin gauge sheet aluminum and other lightweight metals. The feasibility of such uses has been demonstrated for closure panels in both helicopters and aircraft. Thus, an access panel tested for use on a fighter aircraft is shown in Fig. 9.16, where 1.6 mm inner and outer 7075 T6 aluminum skins have been ultrasonically welded into a panel of approximately $0.8\,m \times 0.6\,m$.

9.6 Summary of process advantages and disadvantages

Having reviewed the principles, key features and applications of ultrasonic metal welding, both the advantages and disadvantages of the process may be evident at this stage. Nevertheless, a summary of these features is in order.

Advantages

- A solid-state welding process – hence low heat
- Excellent for Al, Cu and other highly thermal conductive materials
- Able to join wide range of dissimilar materials

- Can weld thin–thick combinations
- Welds through oxides and contaminants
- Fast, easily automated
- No filler metals, welding gases required
- Low energy requirements.

Disadvantages

- Restricted to lap joints
- Limited joint thickness
- High strength, high hardness materials difficult to weld
- Material deformation may occur
- Noise from part resonance may occur
- The process is often unfamiliar.

These points will be discussed below in more detail.

9.6.1 Advantages

Solid-state joining process

Many of the advantages of ultrasonic metal welding stem from its basic solid-state nature. The weld zone, where joining of parts occurs, is a region of plastic deformation and flow, resulting in local mechanical distortion of grain structure, but with absence of melting, fusion, or other evidence of high temperatures relative to melting. Little modification of grain structure is seen away from the deformation zone, which is itself thin, of the order of tens of micrometers at most.

Welding of aluminum, copper, etc.

The ultrasonic materials weldability matrix shows 'in principle' that nearly every material and most material combinations can be welded. Nevertheless, the softer metal alloys, such as those of aluminum and copper have been shown to be the most weldable. Because of the high thermal conductivity of these alloys, they have often proved the more difficult to weld by more traditional methods, such as resistance spot welding. This benefit of welding high thermal conductivity materials is, of course, tied to the solid-state, non-thermal nature of the process.

Joining of dissimilar materials

Note has been taken of the range of dissimilar metal combinations that are reported to have been welded; this advantageous feature is again due to the solid-state nature of the process. A wide variation in welding conditions, depending on materials may be necessary. Thus, a weld readily achieved with 1.5 kW, 0.25 s and 900 N of static force in 1 mm 6061T6 aluminum, may require 2.5 kW, 0.5 s and 1800 N for 2.5 mm, if the weld is to be made. In general, well-developed weld procedures for a wide range of materials and conditions do not currently exist (as they do, for example, in arc welding), so that each combination must be approached in an exploratory manner. In some cases, joints in widely dissimilar materials are assisted by thin foil transition layers of some intermediate material. Joints have also been made in metal–ceramic, metal–glass combinations, in some cases using transition layers or metallized coatings.

Thin–thick combinations

Yet another advantage accruing from the solid-state nature of the process is the ability to join thin sheets and foils (in the ranges of 0.025–0.250 mm) to thick parts. Making such joints, especially in materials with high thermal conductivity can be difficult using other procedures due to the large heat sink of the thick material drawing heat from the weld zone. Such joints are readily made with ultrasonic welding, including joining of multiple foils to a thick substrate with a single weld.

Oxides and contaminants

By the very nature of the inherently friction-like ultrasonic welding action, oxides and contaminants are fractured, disrupted and dispersed. For example, any oil-like surface residues are quickly vaporized in the early stages of the cycle, and some oxides will be dispersed into the weld periphery, or into the turbulent micro-volumes of plastically deformed material at the interface. In summary, the ultrasonic welding process can be tolerant of less than ideal surface conditions. Nevertheless, there is sometimes the tendency to assume the process can achieve successful welds through *any* level of surface contamination, or that control of surface conditions is unnecessary, both of which are unsound assumptions. As a minimum, different conditions will yield variations in the process, such as changeable weld times, and as a maximum, variation in a weld quality such as strength. While special cleaning methods may only be rarely needed, due consideration must be given to consistency of surface conditions.

Fast, easily automated

The typical ultrasonic weld cycle, from start to stop of the ultrasonic vibration, is a fraction of a second, with 0.2–0.5 s being common. Other time components of the weld-to-weld cycle are tied to details of the welder clamping force and retraction mechanism, and are thus subject to being minimized by appropriate machine design and control. Any other aspects of weld-to-weld time are a function of the manufacturing system requirements of the specific application. No demands, such as cooling time or setting time, are placed on total cycle time by the weld process. The basic electromechanical nature of the process and simple mechanical features of the clamping action make it ideal for automation and continued monitoring and control for production and quality assurance.

No filler metals, gases required

Neither filler metals nor shielding nor consumable gases are needed in the usual ultrasonic welding processes. In some cases, in attempting difficult materials combinations, a thin foil transition layer has been placed between the parts. In those cases where welds are made under hazardous conditions (e.g. closure of explosive containers), an inert gas-filled enclosure can be used.

Low energy requirements

Typically, selection of the ultrasonic welding process is based on one or more of the above advantages, most being related to the solid-state nature of the process. If energy use is a factor, it is found that ultrasonic welding is a low energy user compared to other processes; thus, it uses about one-sixth of the energy of resistance spot welding for comparable welds and even less if compared to arc process welding.

9.6.2 Disadvantages

In considering some of the process disadvantages, we should note the following:

Restricted to lap joints

The requirement to achieve a friction-like interface vibration and the limits on the amount of mass that can be moved with ultrasonic systems, logically requires the moving welding tool to be close to the interface, which results in the lap joint constraint of the process. Thus, butt, tee and corner joints, easily made in fusion processes, are not yet possible with current ultrasonic systems.

Limited in joint thickness, material hardness

These items have been covered in some detail and appear in both cases to be related to current restrictions on available ultrasonic power. In aluminum alloys of the 5XXX, 6XXX classes, 2 mm joints are close to the outer limits achievable with current 20 kHz systems and drop to the order of 0.1 mm for titaniums and harder alloys.

Material deformation

The ultrasonic weld tip will typically create some deformation of the top surface in the softer alloys. This will be more or less pronounced depending on weld tip surface (whether flat, slightly hemispherical and/or serrated or otherwise roughened or knurled) and welding conditions (forces and powers involved). Deformations depths of 5% to 10% of part thickness may result. Anvil-side deformations can arise from the same circumstances, but typically are far less pronounced. Through special attention to tip design and parameters, it is usually possible to reduce, but seldom eliminate, some part deformation or marking.

Noise

Being an 'ultrasonic' process, it would seem that audible sound would not be an issue in ultrasonic welding. Two aspects arise, however. The first and most common is that the 20 kHz (or higher) welding frequency may induce subharmonic vibrations in larger parts, which are in the audible range. In such cases, it may be possible to dampen these modes by light clamping of the part at one or more locations. The second aspect is that for some higher power welders, a driving frequency of 15kHz–16kHz may be used, frequencies that are in the high audible range. For these cases, an acoustic enclosure is needed to shield the radiated sound.

Process unfamiliarity

The very large majority of welding processes used in production are fusion based, using electric arcs, resistance heating or high energy beams, such as lasers or electron beams. Ultrasonic metal welding, being a solid-state process, and, further, one that is based on high frequency vibration mechanics, is not typically encountered in the education and experience of manufacturing and welding engineers. Dealing with the unfamiliar is not typically a comfortable option, especially when faced with the pressures of modern manufacturing, and can be a disadvantage in considering the ultrasonic process.

9.7 Future trends

The future direction of ultrasonic metal welding is being driven by pushing back the current main boundaries of the process which are (1) joint thicknesses, (2) weldable materials and (3) joint types. The steps being taken in these directions, or that can be envisioned, are briefly summarized below.

9.7.1 More powerful welding systems

Most ultrasonic metal welders operate at 20 kHz and in the 2.5 kW–3.5 kW range. Such systems have been sufficient to achieve most of the results described previously, but also face limitations in joint thicknesses and materials that can be joined, as have been described. One area of future development will be the introduction of more powerful welding systems, with this achieved by development of more powerful transducers. Thus, 5 kW–6 kW systems operating at 20 kHz are now becoming available on a select basis. This trend is expected to continue, with powers climbing to yet higher levels, potentially to 10 kW. An additional way of increasing power of welders is to couple two or more additional transducers to the weld head, achieving an additive effect of the individual transducers. An example of this was shown for the case of the torsion welder, Fig. 9.9, where four transducers were harnessed in a push–pull fashion. The design of coupling devices requires the solution of a complex tool vibration problem, but has been done for special cases. Another alternative being explored is to apply vibrations to both top and bottom of the workpieces, using two welding systems.

9.7.2 Mechanism of ultrasonic welding

Over the years, research has significantly advanced understanding of the ultrasonic metal welding process, both from a metallurgical and a mechanics perspective. Still, much remains to be done in several areas; for instance, the full range of the weldability of materials must be better understood, including specific procedure data on the various metallurgical combinations. This latter relates to development of benchmark welding procedures for some of the more common material combinations. A necessary and important development is that of providing a mechanics-based model of the welding process, with such a model relating welding and weld quality (e.g. weld strength) to measurable weld parameters, such as vibration amplitudes, forces and power inputs. Some work has been accomplished here (e.g. Gao and Doumanidis, 2002; de Vries, 2004), but more remains to be done in order to provide methods of real-time monitoring and quality assurance.

9.7.3 Joint types

Some limited progress has occurred in this area, but progress may be slow in achieving welds in completely new joints. Achievement of butt welds has been reported, for example by Tsujino *et al.* (2002), who has developed a means of creating a butt weld by clamping one part and creating a bending resonance in the second part, resulting in the friction-like vibration at the interfaces characteristic of the ultrasonic welding action. For small parts, butt welds have been made by attaching the small part to the vibrating weld tip. These developments notwithstanding, it is difficult to envision ultrasonic butt welds becoming routine joint geometry. One may also inquire whether a concept of ultrasonic metal 'far field' welding might develop, such as is done in ultrasonic plastic welding, where vibrations are applied to a part at one location and a weld created at a part interface some distance removed. If such were possible for metals, butt and tee joints would be feasible. However, no instances of such welds are known. It is believed that joint types for ultrasonic metal welding will largely remain restricted to lap-type configurations.

In summary, of the three areas of (1) joint thicknesses, (2) weldable materials and (3) joint types, significant progress is expected in the first two, driven by more powerful welding systems and improved understanding of the ultrasonic welding mechanism, but with the process still having its main application on lap-type joints.

9.8 Sources of further information and advice

American Welding Society, (1991). 'Ultrasonic welding', *Welding Handbook*, 8th ed., v. 2, Miami FL, American Welding Society

Baladin G., Kuznetsov V. and Silin L., (1967). 'Fretting action between members in the ultrasonic welding of metals', *Welding Production*, **10**, 77–80

Beyer W., (1969). 'The bonding process in the ultrasonic welding of metals', *Schweisstechnik*, **19**(1), 16–20

Chang U.I. and Frisch J., (1974). 'On optimization of some parameters in ultrasonic metal welding', *Welding Journal*, **53**(1) 24–35

Harthoorn J., 'Joint formation in ultrasonic welding compared with fretting phenomena for aluminum', *Ultrasonics International, Conference Proceedings*, (1973). Guildford, UK, IPC Science and Technology Press, 43–51

Hazlett T. and Ambekar S., (1970). 'Additional studies on interface temperatures and bonding mechanism of ultrasonic welds', *Welding Journal*, **49**(5) 196s–200s

Heymann E. and Pusch G., (1969). 'Contribution to the study of the role of recrystallisation in the formation of the joint in ultrasonic welding', *Schweisstechnik*, **19**(12) 542–5

Jones J.B., Maropis N., Thomas J.G. and Bancroft D., (1961). 'Phenomenological considerations in ultrasonic welding', *Welding Journal*, **40** 289s–305s

Joshi K., (1971). 'The formation of ultrasonic bonds between metals', *Welding Journal*, **50**(12) 840–8

Kreye H., (1977). 'Melting phenomena in solid state welding processes', *Welding Journal*, **56**(5) 154–8

Kreye H. and Wittkamp I., (1975). 'On the bonding mechanism in ultrasonic spot welding', *Schweissen Schneiden*, **27**(3) 97–100

Mitskevich A., (1973). 'Ultrasonic welding of metals', in Rozenberg L.D. (ed.) *Physical Principles of Ultrasonic Technology*, v.1, Part 2, New York, Plenum Press

Neppiras E., (1965). 'Ultrasonic welding of metals', *Ultrasonics*, **3** 128–35 (Jul–Sep.)
Neville S., (1961). 'Ultrasonic welding', *British Welding Journal*, 177–87 (Apr.)
Okada M., Shin S. and Miyagi M., (1963). 'Joint mechanism of ultrasonic welding', *Japan Institute of Metals*, **4** 250–6
Pfluger A. and Sideris X., (1975). 'New developments in ultrasonic welding', *Sampe Quarterly*, **7**(1) 9–19
Reuter M. and Roeder E., (1993). 'Ultrasonic welding of glass and glass-ceramics to metal', *Schweissen Schneiden*, **45**(4) E62–E65
Wagner J., Schlicker U. and Eifler D., (1998). 'Bond formation during the ultrasonic welding of ceramic with metal', *Schweissen Schneiden*, **50**(10) 636, 638, 640–2
Weare N., Antonevich J. and Monroe R., (1960). 'Fundamental studies of ultrasonic welding', *Welding Journal*, **39**(8) 331s–341s
Wodara J., (1986). 'Joint formation in the ultrasonic welding of metallic substances', *ZIS Mitteilungen*, **28**(1) 102–8
Wodara J., (1986). 'Ultrasonic weldability of metals', *ZIS Mitteilungen*, **28**(2) 230–36
Wodara J. and Eckhardt S., (1982). 'Determination of temperature fields in the ultrasonic welding of metals', *Schweisstechnik*, **32**(10) 436–7
Wodara J. and Sporkenbach D., (1989). 'Exploiting the frictional processes occurring during ultrasonic welding to improve the weldability of metallic materials', *Welding International*, **3**(5) 450–3

9.9 References

de Vries E., (2000). *Development of the Ultrasonic Welding Process for Stamped 6000 Series Aluminum*', Diploma Thesis, University of Applied Science, Emden, Germany
de Vries E., (2004). *Mechanics and Mechanisms of Ultrasonic Metal Welding*, PhD Dissertation, The Ohio State University
Gao Y. and Doumanidis C., (2002). 'Mechanical analysis of ultrasonic bonding for rapid prototyping', *J. of Manufacturing Science and Engineering*, **124**, 426–34
Harthoorn J., (1978). *Ultrasonic Metal Welding*, PhD Dissertation, Technical University Eindhoven
Vitek J. and Miklanek L., (1978). 'Technological requirements for quality assurance at the ultrasonic welding of metals', *Schweisstechnik*, **28**(7) 316–7
Tsujino J., Hidai K., Hasegawa A. *et al.* (2002). 'Ultrasonic butt welding of aluminum, aluminum alloy and stainless steel plate specimens', *Ultrasonics*, **40**(1–8), 371–4

10
Occupational health and safety

F. J. BLUNT, University of Cambridge, UK

10.1 Introduction

The welding industry is a major player in manufacturing. It encompasses the traditional arc and gas processes as well as advanced techniques such as laser welding, friction welding and electron beam welding. More innovative use of materials leads industry to a need to find techniques to join them and the more advanced welding processes will often fulfil that role. New materials can potentially bring new hazards into the workplace.

In 1998 a group of senior managers and experts from the welding community met at a 'Vision Workshop' to look ahead to where the industry might be in 2020. The report of the workshop proceedings contains two items of relevance to this chapter; first, one of the strategic goals in relation to the environment was to reduce energy use by 50% by reducing pre- and post-heating operations and through the use of lower heat input welding processes. Second, the report speaks of a vision of a workplace with improved conditions for the workforce, dispelling the image of welding as 'dark, dirty and dangerous'.[1]

This chapter looks first at legislative changes relating to health and safety in Europe. Readers outside Europe may also find this section informative, since the constraints set in Europe are based on the same research data that are available to all. There is continual pressure to reduce the incidence of disease related to substances hazardous to health. Of particular interest to welders is the effect of the various constituents of fume, since many welding processes, by their nature, will always produce fume. Planned future legislation aims to reduce risks to the workforce in the areas of vibration and noise. Recent legislation in Europe has clarified the control measures that are to be expected in workplaces that have the potential to contain explosive atmospheres.

The chapter will then summarise some of the scientific research that has recently been carried out. This includes some work on explosion risks in preheating. There is some work on the measurement of fume using an improved capture method. Ongoing research aims to improve our fundamental understanding in two areas – exposure to vibration and exposure to electric and magnetic fields. Some of the environmental issues that are currently of global significance are described and their present and likely future effects on the welding industry are reviewed. The chapter ends with some sources of further information and advice. The section includes sources of legislation, the enforcement agencies and some of the key government agencies, research bodies and national and international organisations. This chapter does not give a general overview of health and safety in welding. This role is fulfilled by the book *Health and Safety in Welding and Allied Processes*, 5th ed.,[2] which contains both an overview of and specific guidance for the major welding processes, both for readers in the UK and readers in the USA.

10.2 Legislation

10.2.1 Legislative drivers in Europe

For countries within the European Community a significant amount of the law concerning industrial matters emanates from Directives that are enacted by the European Parliament and Council. Some of these Directives seek to establish freedom of trade within Europe and are concerned with setting minimum agreed standards for manufactured goods. This allows products, including machinery, to be marketed freely within Europe. Other Directives directly concern health and safety at work. They do not automatically become law in the member states, but are implemented within each state using their own legislature, to a timetable that is set by the European Parliament. Directives tend to be goal-setting rather than prescriptive. Since it takes many years for a European Directive to be enacted within member states it allows those who are to be affected by it to contribute to the consultation process and influence any decisions that are to be made in implementing it.

10.2.2 Hazards from fume

Many of the constituents of fume are known to have adverse effects on health. The constituents can include a wide range of metallic and non-metallic elements, oxides and other compounds. There are overall exposure limits on welding fume and there are individual exposure limits on many of the constituents of welding fume. Recently the exposure limit for manganese and its inorganic compounds (which are present in fume from manganese-containing materials) has been under review in the UK and it has now been reduced[3] from an occupational exposure standard of $1\,mg/m^3$ to a workplace exposure limit of $0.5\,mg/m^3$.

Table 10.1 summarises the UK workplace exposure limits to those substances that a welder may encounter in the workplace.

272　New developments in advanced welding

Table 10.1 Workplace exposure limits for substances commonly found in welding and allied processes (Source: *Workplace Exposure Limits*[3])

Substance	Limits based on an 8-hour time-weighted average		Limits based on a 15 minute time-weighted average	
	ppm	mg/m^3	ppm	mg/m^3
Cadmium oxide fume	–	0.025	–	0.05
Cobalt and compounds	–	0.1	–	–
Chromium VI	–	0.05	–	–
Manganese and its inorganic compounds	–	0.5	–	–
Nickel and its compounds	–	0.1 (soluble) 0.5 (insoluble)	–	–
Trichloroethylene	100	550	150	820

In the UK the Health and Safety Executive has set a target for the reduction of occupational asthma by 30% by 2010. Occupational asthma is a term that is specifically used to describe a condition where exposure to a substance at work produces a hypersensitive state in the worker's airways, and it triggers a subsequent reaction in them. This is a form of allergic reaction and in the worst cases can lead to a person having to change their job altogether. There are significant numbers of cases of occupational asthma reported among welders.[4] While welding fume is not among the top eight agents that cause occupational asthma in the UK, nevertheless Government statistics indicate that the welding trades have the fourth highest incidence of occupational asthma.

Stainless steel is implicated in many cases of occupational asthma in welders, and it would be prudent to ensure that the local exhaust ventilation used when welding stainless steel is in excellent working order. General ventilation would not be considered adequate in controlling exposure to an asthmagen. The employer should be aware of the possibility of workers becoming sensitised and have a health surveillance programme that can identify the early symptoms. Workers should be given information about the hazards and shown how to minimise the risk to themselves. They should be told about the early warning signs, which may include coughing, wheezing and chest tightness, a runny or stuffy nose and watery or prickly eyes.

10.2.3 Work in potentially explosive atmospheres

Guidance for hot work in potentially explosive atmospheres, such as repair of petrol tanks, has been in existence for a very long while. Legislation has also been in place to lay down minimum standards for the storage and use of highly flammable liquids and liquefied petroleum gas.

As a result of the implementation of the ATEX Directive[5] legislation is now in place across Europe to protect workers from dangerous substances (explosive, oxidising, extremely flammable, highly flammable or flammable) and potentially explosive atmospheres. In the UK this has been implemented as the *Dangerous Substances and Explosive Atmospheres Regulations* 2002.[6] These repeal the former legislation relating to highly flammable liquids and specifically require employers to carry out risk assessment and implement control measures for work with dangerous substances and in potentially explosive atmospheres.

A dedicated code of practice is planned for welding operations on containers that have previously contained materials that might cause explosion. The first priority is to consider whether it is feasible to do the work using a method that does not generate heat or sparks. If hot work does prove necessary it must be carefully planned. The draft code of practice formalises the requirement for work to be done under a permit-to-work, for adequate cleaning, inspection, monitoring and control.

When planning welding in a confined space, employers should specifically consider the possibility of leaks of oxygen and the practicality of odorising the oxygen if leaks are possible. They should consider the possibility of fires and explosions due to flashback, decomposition of acetylene, and high-pressure oxygen. They should plan for the safe storage of gas and how to avoid the spread of fire to other combustible materials.

Many of the requirements that are formalised by this legislation are already considered to be good practice in the welding industry and are described in various publications.[7–12] However, one aspect of the new legislation that will need careful consideration by employers is the requirement to zone areas where gases or solids that can form explosive atmospheres could be released. The process of zoning leads to a specification of what electrical equipment will be allowed in the area. Formal zoning is a new requirement and it would be expected that at the very least the gas stores should be zoned, to define the area around them that should be free from sources of ignition.

10.2.4 Vibration

It has been known since 1911 that persistent use of certain types of vibrating tools can eventually lead to permanent damage to the hands. Early indications of damage are a numbing and blanching of the fingers and this is the origin of the name 'vibration white finger'. If damage continues it may result in irreversible changes in the nerves, muscles, bones and joints. Welders are potentially exposed to hand–arm vibration due to their use of tools such as hand grinders, chipping hammers and needle guns.

Vibration is measured in a similar manner to noise – indeed some noise meters can also measure vibration. It is believed that the following factors are important in characterising vibration exposure that may be harmful:

- magnitude, frequency of the vibration, the total daily exposure and the pattern of exposure and rest periods;
- cumulative exposure over the worker's lifetime;
- grip or the force that the user applies to the vibrating tool and their posture;
- area and the part of the hand that is in contact with the vibrating tool;
- type of tool and the type of workpiece;
- susceptibility of the individual, which includes such factors as smoking;
- climate.

However, even knowing these factors, it is not yet possible to predict the likelihood of vibration damage, neither is it possible to detect it in its very early stages. The National Institute for Occupational Safety and Health (NIOSH) has an informative criteria document[13] that describes the condition and recommends how employers should avoid vibration-induced damage. This document does not set exposure limits, but recommends proactive measures such as medical monitoring and surveillance, engineering controls, good working practices, use of protective clothing and equipment, worker training programmes and administrative controls such as restricting the hours of use of such tools. It has a useful review of the standards and recommendations from other national organisations current at its date of publication (1989).

NIOSH is currently engaged in research projects[14] to try to establish stronger links between cause and effect. They hope to use microscopy to examine the capillaries at the base of the fingernail, to see whether they can predict adverse effects from physical changes there. Using computer models of stress and strain they hope to be able to relate the way in which the soft tissues of the hand are compressed and displaced and use these as a way to predict adverse effects.

Those who already have vibration white finger are known to experience delays in the return of the warmth to their fingertips after exposure to cold. It is hoped to use infrared imaging to monitor this and use it to assess the severity of the condition. NIOSH researchers hope to instrument a chipping hammer to be able to measure the impulse at its tip simultaneously with the vibration in the handle. In this way they hope to be able to monitor the effectiveness of anti-vibration methods while being able to see whether the effectiveness of the tool remains the same. A poorly designed 'low vibration' tool may be less effective than its higher vibration counterparts, which in turn may lead to longer periods of use and the benefits of 'low vibration' may be entirely lost. Researchers will also measure the effectiveness of anti-vibration gloves by using an instrumented vibrating handle that can simulate various vibrating tools.

While clearly not all the characteristics of the damaging effects of vibration are known, action levels have been identified for whole-body vibration and hand–arm vibration to safeguard the health of workers. Vibration is expressed as an acceleration, in m/s^2, since the degree of harm that a human suffers is related to the acceleration. The measurement is averaged over an 8 hour period, representing a nominal day's work, and is measured as a function of frequency. This is then weighted, since the response of the body is known to be different at different vibration frequencies, the most important being in the range 5–20 Hz, and the weighted average is designated an A(8).

There is new legislation in the UK specifying maximum vibration exposure. Formerly the advisory limits on hand–arm vibration in the UK were based on the British Standard BS6842, 1987, which has now been withdrawn. This limit was 2.8 m/s^2 A(8), calculated from the magnitude of vibration in the dominant axis. There is a European Directive 2002/44/EC, on the subject of vibration,[16] which has been implemented by Control of Vibration at Work Regulation 2005[15]. The new legislation uses vibration measurements carried out in accordance with the new Standard BS EN ISO 5349-1, 2001,[17] which calculates the vibration magnitude using measurements in three directions. Vibration magnitudes calculated in this way are larger than those obtained using the old standard by a factor of between 1 and 1.7. The new legislation defines two new exposure levels, an action level at 2.5 m/s^2 A(8) and an exposure limit of 5 m/s^2 A(8). It will require employers to reduce hand–arm exposure to a minimum, provide information and training, assess exposure levels, carry out a programme of measures to reduce exposure and provide appropriate health surveillance when exposure reaches the exposure action level. There is a requirement to keep exposure to below the exposure limit except under certain specified circumstances.

In the workplace, the employer can reduce the incidence of vibration-induced disease by automation and mechanisation, by purchasing low vibration tools, by reducing exposure times, by maintaining the equipment in efficient working order and by giving instruction in correct operating techniques. The workers should be educated to recognise the symptoms – that numbness or tingling after using vibrating tools may be an early warning sign and should be reported to their supervisor. Employees can help to reduce the risks by keeping warm, by avoiding smoking and by taking exercise.

10.2.5 Noise

Several welding, cutting and gouging processes are noisy, to the extent of exceeding the thresholds currently specified in UK and USA legislation. Examples include:

- Gouging, which can produce noise levels over 90 dB(A)
- MIG (metal inert gas) (GMAW, gas metal arc welding) welding which can exceed 90 dB(A)

- Plasma cutting, which can produce noise levels of the order of 110 dB(A).

Some processes associated with welding, such as grinding, can also produce extremely high noise levels. Our susceptibility to damage by noise, like our susceptibility to vibration, depends on frequency and the measurements are weighted to reflect the sensitivity of the human ear. Weighted measurements are denoted dB(A) in the UK, and dBA in the USA, and for simplicity dB(A) is used for the rest of this section.

Current legislation in the UK[18] and USA[19] (2005) require measures to be taken to protect the hearing of workers when noise levels reach 85 dB(A) averaged over an 8 hour working day, and demand protective measures to be taken to safeguard their employees' hearing. The wearing of hearing protection becomes mandatory in the UK at a threshold of 90 dB(A). Currently in some European countries the thresholds at which employers must take action are already lower than these figures.

The exposure limits currently (2005) used in the USA and UK represent thresholds at which there is a given probability of hearing damage – they are not thresholds of safety. Research has indicated[20] that approximately 5% of workers exposed to 90 dB(A) for an 8 hour period daily will experience a 30 dB hearing loss at 1, 2 and 3 kHz after 25 years, rising to almost 20% after 45 years. The corresponding figures for exposure to 85 dB(A) are approximately 2% and 10%. Therefore, even when the current exposure limits are applied, there will be a number of individuals who will experience hearing loss as a result of their work. The hearing loss that is suffered as a result of exposure to excessive noise tends to be in the frequencies that are necessary for the clear understanding of speech and the loss cannot be compensated for by a hearing aid.

The European Community has passed a Noise Directive (2003/10/EC)[21] that is on a timetable for implementation in member states by 2006. When implemented, this legislation will, like previous legislation, require employers to assess noise levels where workers are likely to be exposed to risks, eliminate risks at source or reduce them to a minimum and implement appropriate health surveillance where the risk assessment indicates a risk to health. In justified circumstances weekly averaging of noise will be permitted, instead of using an 8 hour averaging period.

The most noticeable changes in the UK introduced by the Control of Noise at Work Regulations 2005[22] are the new exposure thresholds coming intro force in April 2006, which are significantly lower than the current figures. The new limit on personal noise exposure is 87 dB(A) and 140 dB (C-weighted) at the ear. Where personal exposure, not taking hearing protection into account, exceeds 85 dB(A) and 137 dB (c-weighted), there is a requirement to establish and implement a programme of technical and/or organisational measures to reduce exposure to noise. Areas where noise levels exceed

85 dB(A) and 137dB (c-weighted) will need to be marked, with access restricted where technically feasible and where the risk of exposure justifies it. This threshold also triggers mandatory use of hearing protection and appropriate health surveillance Where exposure, not taking hearing protection into account, exceeds 80 dB(A) and 135 dB (c-weighted) hearing protection must be made available, information and training must be provided and audiometric testing provided where a risk to health is indicated.

10.3   Recent and ongoing research

10.3.1   Fundamental difficulties

Despite the labour figures indicating that around 400 000 people in the USA are directly engaged in welding, it is difficult to research health effects and make positive associations between causative factors and those effects. Working environments are complex, and each of us is exposed to a wide range of both physical and chemical agents at work. We have different lifestyles, different diets and different susceptibilities to disease. All of these factors make it difficult to pinpoint patterns of exposure and associate them with particular health effects – especially if the association is only weak. It is rarely acceptable to embark on experimentation with human subjects.

NIOSH recently published a comprehensive review[23] noting that past research discovered that various respiratory disorders are found in large numbers of welders. It is known that nickel and chromium (VI) are classified as carcinogens and that chronic exposure to manganese has been associated with a disease similar to Parkinson's. However, we do not have data to indicate whether welders are exposed to these substances in such quantities that they could trigger these effects, or how such exposure can lead to serious long-term effects.

Many past studies have involved comparing large groups of people to try to associate patterns of disease or causes of death with differences in their exposures and lifestyles. This is not always a successful strategy, due to many problems. Such studies often use death certificates to identify the diseases that people suffered. However, death certificates generally only record the immediate cause of death and may not record the underlying cause(s) of death, or they may not mention a condition that was present, but would have been of interest to the research study.

In retrospect, it is difficult to quantify exposure of individuals to different agents unless measurements have been taken during their working life. Making such measurements presupposes that the things that must be measured are already known to be the key factors in the development of the disease. Subsequent assumptions about exposures may be wrong – and an example of the problems with making assumptions using generic job titles is evident in the more recent research in electric and magnetic fields, outlined in Section 10.3.4.

Research into health effects therefore tends to be an iterative process where a prevalence of a particular disease is noted to be associated with a particular exposure or a particular trade. Subsequent research attempts to show that it is specific to a certain agent and attempts to rule out other influences, such as bias in the choice of age of the subjects, coincidence, the presence of another agent that has not been taken into account, etc. The data we have at present are too limited and further research is necessary.[24] We need a continuation of epidemiological studies[24] – investigating the patterns of disease among populations of interest. This is needed to gain a better understanding of the role that welding fume may play in the suppression of the immune system, the development of lung cancers, neurotoxicity, skin damage, reproductive disorders and the other effects that prior studies have associated with the components of welding fume. The second strand of research is at the molecular level[24] to gain an insight into the ways that changes in cells or in genetic material can lead to tumour formation, nerve damage or other adverse changes.

10.3.2 Fume measurements

Welding, by its nature, has many variables – among them are the process itself, the consumable (where applicable), the parent metal, flux and/or shielding gas (where applicable), voltage, current and standoff. Fume emissions depend on all the variables. While there are many fume emission measurements relating to the various arc welding processes, research continues. Continued research is important in order to produce high quality, reliable and reproducible data against which one can formulate control measures to safeguard health and which can be used to verify mathematical models.

Recently a research group has devised an improved design of welding chamber for the capture of fume for analysis,[25] which has allowed more reproducible and accurate measurements to be made. This enabled them to map out the fume emission rates for GMAW, as a function of voltage and current, with much greater precision than was previously achievable. They found a complex relationship between fume emission rate and welding parameters. Fume emissions rose as the current, voltage and wire feed speed increased in globular transfer mode, only to drop suddenly when the mode changed to spray transfer. At even higher voltages the fume emission rates increased once more.

Fume emission measurements are required wherever new materials are welded, or a new consumable is developed. They are also needed whenever a significant change in working practice is made that might have a bearing on the quantity or identity of the emissions – for instance when welding through coatings.

It is common practice in the automotive industry to weld materials that have been treated with sealants, or have adhesives on their surfaces. Resistance welds are routinely made through these materials and fume is generated as a result. A recent research project was carried out at The Welding Institute (TWI)[26] in which resistance welds were made on pieces of metal coated with a range of typical sealants and adhesives, representing the most widely used types in the industry. Emissions included benzene, 1,3-butadiene and several other compounds, but the concentrations were low in comparison with the total welding fume.

10.3.3 Dangers of explosion

It has long been known that hot work on tanks that have contained flammable liquids requires special measures to ensure that explosion will not occur. However, a less well-known cause of explosion, the ignition of unburnt gas during preheating of weld preparations, has recently been the subject of research.[27] Preheating is commonly carried out using a propane torch, where the flame contains regions of unburnt gas. Under certain circumstances unburnt propane can pass through the welding gap. If the space behind is confined, subsequent explosions are possible and fatal accidents have occurred.

The research indicated that unburnt gas passing through the welding gap collects in the space behind where there is, during the time of preheating, insufficient oxygen to cause ignition. However, when the flame is removed the weld preparation begins to cool and air is drawn into the space. If subsequently an ignition source is brought to the gap the unburnt gas behind it may explode. Both large confined spaces, such as legs for oil platforms, and small confined spaces are susceptible. To avoid this sequence of events the recommendation is, for preference, to avoid the fabrication of an enclosure by welding. Alternatively, other forms of preheating are recommended in place of the use of a fuel gas. Possibilities include induction heating or radiant gas burners.

If a gas torch is to be used welders are cautioned:

- to light the torch correctly by pointing it downwards towards a horizontal surface to trap the vapour and to light it quickly;
- to avoid damaging the nozzle, and check that there is no leakage of fuel from the rear of the heads;
- to choose a welding gap that is 5 mm or more;
- to use a standoff distance that is as large as practicable – at least 150 mm;
- wherever reasonably practicable, the space behind the preparation should be checked with a flammable gas detector after preheating and before bringing another ignition source up to it;
- ventilation behind the gap should be maintained where reasonably

practicable to keep the unburnt gas to below 10% of its lower explosive limit;
- routinely to check hoses, regulators, flame arrestors for integrity.

The British Compressed Gases Association (BCGA, United Kingdom)[28] and the Compressed Gas Association (CGA, United States of America)[29] both publish documents that give advice on the selection, use and maintenance of hoses, regulators, gas torches and other such equipment.

10.3.4 Electric and magnetic fields (EMF)

Public concern continues to grow over exposure to electric and magnetic fields. This has at least in part been fuelled by the rapid increase in mobile communications, with its associated transmitters, and hand-held telephony equipment. We are all exposed to both electric and magnetic fields. Magnetic fields are generated by the passage of an electric current and are therefore larger close to electrical equipment drawing relatively large currents such as sewing machines, magnetic resonance imaging machines, computers and can openers. Magnetic fields generally decay very rapidly with distance from electrical appliances, but are difficult to shield. They are only present when the equipment is actually energised and working. The electric field is generated by the voltage between an appliance or a cable and earth. Electric fields do not decay, but are easily shielded by objects that conduct electricity, which includes buildings and trees.

There has been much research into the effects of both electric and magnetic fields. The subject is complex, because the effects, if any, may depend on the frequency of the field, the strength, whether it is electric, magnetic or both, and the peak exposures. People in specific age groups may have increased susceptibility – for example children. The research results have so far shown no clear unequivocal evidence for a link between exposure to EMF and adverse health effects. Some of the research has been confounded by a lack of measurements of exposure – early research used job titles to assign workers to low or high electric and magnetic fields. Actual measurements show that this is not likely to have been a reliable indicator. Some typical measurements are given in Table 10.2, showing that the designation 'electrical worker' does not necessarily indicate that the exposure to that individual is greater than in other occupations.[30] Further research is clearly needed. At the time of writing there are around 200 research projects in at least 27 countries. A review of the current status can be found in a document produced by the National Institute of Environmental Health Sciences[30] and there is a great deal of information available from the National Radiological Protection Board (NRPB).[31]

The welder is potentially exposed to both magnetic and electric fields, but the magnetic field is believed to be the more significant as it is slightly

Table 10.2 A selection of average exposures of various workers to magnetic fields

Type of worker	Average daily exposures/ mG*	
	Median	Range
Clerical workers with computers	1.2	0.5–4.5
Machinists	1.9	0.6–27.6
Electricians	5.4	0.8–34.0
TV repairers	4.3	0.6–8.6
Welders	8.2	1.7–96.0
Sewing machine operators (Finland)	22.0	10.0–40.0

*1 mG is equal to 0.1 µT (microTesla).
(Source: National Radiological Protection Board[31])

elevated compared to that in many other occupations, as shown by the figures in Table 10.2. At the current level of knowledge there is no proven link between exposures at the levels experienced by welders to adverse health effects. However, since the research is inconclusive, in line with the precautionary principle it is suggested that welders do not expose themselves unnecessarily to magnetic fields. This can be done by welders avoiding wrapping the cable around their bodies and by keeping the welding cable and the return cable close together.

Workers with medical prostheses are a special group. There is a possibility that workers wearing certain types of pacemaker, for certain heart conditions, may be adversely affected by the rather large fields generated by a resistance welding machine. It is recommended that the advice of the consultant physician who is managing the worker's heart condition is sought if they are worried, or if their job brings them close to high magnetic fields.

The legislative timetable relating to electromagnetic fields is not yet fixed. The fall-back position in the UK is that there is the expectation that employers will apply general health and safety legislation to this topic, and refer to the guidelines of the NRPB.[31] There is a European proposal for a Directive on the exposure of workers to electromagnetic fields and waves.[32] The Directive is concerned with the acute effects of electromagnetic fields, which are apparent at relatively high fields. The proposal states that there is as yet no conclusive evidence linking these fields to cancer. Proposed 'action values' and exposure limits are in the document. The action values are the same as those listed in Table 6 of the International Commission on Non-Ionizing Radiation Protection (ICNIRP) guidelines document.[33] The proposed exposure limit values are the same as those listed in Table 4 in the ICNIRP document, for occupational exposure. If the proposed directive is adopted, new regulations will be made. Where the exposure 'action values' are exceeded, employers will be required to put into place an action plan to reduce exposure to a minimum. This will include a consideration of adopting different working methods that entail less exposure, the choice of appropriate equipment, technical measures to reduce the emission of fields, appropriate maintenance, the design and layout of workplaces and workstations, administrative measures, information and training, the limitation of exposure and the availability of adequate personal protective equipment.

10.4 Environmental issues

10.4.1 Introduction

The last 30 or more years have seen a significant awakening of interest in the environment and a much greater understanding of how human activities in one geographical area can have long term and far reaching effects in another. The difficulties in research into environmental effects are possibly even greater than those in epidemiology. It is difficult to obtain measures of changes in the variables in the land, water and air, the effects of such changes on the earth and its climate cannot be predicted, and the effect of changes that might be made to try to reverse a trend are unknown. Changes that are to be made involve co-operation between nations and the involvement of the people within them. They can be difficult to 'sell' because they can be in conflict with economic and social aspirations.

Three drivers in the environment are discussed here:

- Preservation of the ozone layer by restricting emissions of ozone-depleting chemicals;
- Reduction in global warming by restrictions on emissions of greenhouse gases;
- Sustainability in all its forms, which includes the controlled disposal of waste.

In many ways, the third of these drivers incorporates the other two and they are not entirely separable.

10.4.2 Ozone

Ozone is a gas that is formed from oxygen, with the formula O_3. It is relatively unstable, readily reforming oxygen, O_2, especially when it comes into contact with surfaces. Ozone is a familiar gas in the welding workplace, being formed in significant quantities when welding stainless steel and aluminium with the TIG (GTAW) and MIG (GMAW) processes. The mechanism for its formation is the action of ultraviolet (UV) light on oxygen in the atmosphere around the arc. It is not normally found in any significant quantity in processes such as manual metal arc, because of the high levels of fume generated by that process. Ozone is also formed at ground level due to the action of ultraviolet light from the sun on air containing oxides of nitrogen and volatile organic compounds. It is thus found in quite high concentrations in some cities, as a result of pollution from vehicles. Ozone at ground level is a significant hazard to health, as it can cause lung damage.

The 'ozone layer' is a region of the atmosphere high above the earth where a proportion of the oxygen molecules also form ozone due to the action of UV radiation from the sun. In the upper atmosphere the production and persistence of ozone plays a crucial role in filtering out a proportion of the harmful UV coming towards us from the sun, preventing it from reaching the earth's surface. Exposure to UV has a proven adverse effect on humans, being a known cause of skin cancer. Ozone produced at ground level does not persist long enough to drift into the upper atmosphere. The recognition that several substances in widespread use were drifting into the upper atmosphere and reducing the effectiveness of the ozone layer led to the Montreal Protocol in 1987. Chlorofluorocarbons (CFCs) had been identified as ozone depleting substances and an agreement was drawn up to phase out CFCs, along with several other ozone depleting substances, which was signed by around 60 countries.

Worldwide, CFCs were used in aerosols, as solvents, in refrigerants and in foam blowing; their use was widespread. In the welding environment, therefore, the phasing out of these substances has been most noticeable in non-destructive testing, where aerosol dyes and developers are commonplace. Several of the properties of CFCs made themselves attractive for these applications – they appeared largely inert, non-toxic and non-flammable. Substitution for CFCs in aerosols has brought different hazards into the workplace – CFCs are non-flammable, but many of the substitute propellants, such as butane, are highly flammable.

Other substances that have been phased out as a result of the protocol include 1,1,1-trichloroethane and bromochloromethane, which were both marketed under several trade names. These were common solvents for degreasing and, when choosing an alternative, users should assess carefully the potential replacement substance for its toxicity, flammability and environmental effects.

Trichloroethylene is one substitute, but unfortunately this substance suffers from several drawbacks. First, it is a much greater hazard to human health than is 1,1,1-trichloroethane and there have been many instances of people being overcome by entering degreasing tanks when the vapours are present, sometimes with fatal consequences. It has also recently been officially classified as a carcinogen, so those currently using the substance should review their use of it in the light of this reclassification. Note that in Table 10.1 it is listed as having a workplace exposure limit. Users should therefore consider either substitution of a substance with a lower intrinsic level of hazard, or the total enclosure of the processes in which it is used. Many of the old degreasing baths that were used for 1,1,1-trichloroethane do not offer sufficient protection for them to be suitable for use with trichloroethylene.

Large users of solvents will find that their operations fall under the Solvent Emissions Directive,[34] which aims to reduce the quantity of volatile organic solvents being emitted into the atmosphere, particularly those that are designated as carcinogens. Trichloroethylene, as an organohalogen compound and a carcinogen, is on the 'Black List' of substances in the EEC, where the declared intention is to eliminate the pollution they cause.

Volatile organic compounds are also associated with the production of ozone at ground level, as described above. Information on possible substitutes and good practices is available in a leaflet from Envirowise.[35]

Lastly, solvents are generally banned from discharge to groundwater. In particular, even trace quantities of trichloroethylene can render groundwater unusable for drinking.

10.4.3 Global warming

Carbon dioxide and other greenhouse gases such as nitrous oxide, methane and CFCs are implicated in global warming. Measurements going back over a century show a steady increase in the concentration of carbon dioxide in the atmosphere. The sun's radiation, on entering the earth's atmosphere, spans from the UV to the infrared. Some of this is absorbed by the earth's surface and the oceans, some is absorbed by plants and used as a source of energy for photosynthesis and some is absorbed by the atmosphere where it causes changes in pressure which give rise to winds. Energy is reradiated to space with the longer wavelengths (infrared) predominating. However, gases in the atmosphere that absorb the energy in these wavelengths prevent it from being radiated out into space and the energy is retained close to the earth. This leads to the temperature of the earth and its atmosphere being higher than they would be in the absence of this absorption.

While it is true to say that if the atmosphere had no greenhouse gases the earth would be too cold to live on, we are currently concerned that too much carbon dioxide and other gases that absorb infrared radiation may change the climate and make the earth too warm. The model for predicting carbon dioxide levels is imperfect, since it is not yet fully understood where all the carbon is stored in the earth and how the balance of carbon dioxide in the atmosphere is maintained. Stores of carbon include living plants and animals, fossil plants, rocks and a considerable amount of carbon dioxide dissolved in the oceans. Climate change will bring with it changes in the patterns of growth of many organisms, some of which will add to the carbon dioxide in the atmosphere, and some of which will remove it.

Concerns about global warming led to the Kyoto Climate Change Protocol where most industrialised countries agreed to take measures to reduce the emission of substances that contribute to climate change. These substances include CFCs, carbon dioxide, nitrous oxide and methane. The question of CFCs had already been addressed in the Montreal Protocol, due to their ozone depleting potential. Nitrous oxide and methane are produced in large quantities by the decay of biological material. Carbon dioxide is produced by the combustion of carbonaceous materials of all kinds, by decaying organic materials, by respiration and by fermentation. Of these, the release of carbon dioxide from fossil fuels is of most concern, because the carbon in these deposits has been in the earth's crust for millions of years, where it had effectively been removed from circulation.

This driver is leading companies towards consideration of where the greatest emissions of carbon dioxide are produced. A leading manufacturer of welding consumables[36] made an estimate of the CO_2 emissions associated with all its consumables during one year, encompassing the entire life cycle of the consumable from raw material extraction and conversion to its use in welding and disposal of the waste. They have estimated that during the life cycle of the consumable, approximately 41% of the emissions are associated with raw material extraction and conversion, 37% with welding, 11% with production of consumables and 9% in transport. A life cycle analysis is a very powerful tool for assessing impacts and planning for reductions, as it helps to avoid saving in one area only to make the problem worse in another.

10.4.4 Sustainability

Sustainability is generally defined as the ability of the world to meet its needs today without compromising the ability of future generations to do likewise. This includes considerations of the environmental factors already mentioned, but in addition recognises that the earth is a finite resource – there are only fixed quantities of resources such as metals, oil and other fossil fuels. Thus we are concerned with the depletion of resources and the emissions to the atmosphere, water and soil, which result from the fabrication, use or disposal of manufactured articles.

The environmental impact of consumables includes the emissions of gases and particulates, the use of energy and the wastes that are produced. Ultimately, the product that has been fabricated using the consumables will also become waste, unless it is recycled. In the future it is probable that manufacturers will need to invest more effort in designing articles that can be dismantled and reused, or recycled.

Welding equipment itself also has an environmental impact. However, a life-cycle analysis shows that with welding equipment the greatest environmental impact lies in its energy consumption while it is being used. Thus, designing equipment that is energy efficient will have the greatest impact.

10.5 Sources of further information and advice

10.5.1 General advice

The book *Health and Safety in Welding and Allied Processes*[2] describes the key hazards associated with a wide variety of welding processes, the health effects and the control measures that reduce the risks to welders. It addresses the legal requirements of both the UK and the USA and contains almost 200 references.

In major industrialised economies there is a well-established framework governing the obligations of the employer towards the preservation of the health, safety and welfare of his or her employees. Differences are apparent both in the approach that is taken to health and safety, and to the standards that are acceptable. However, there is much common ground, as would be anticipated because the basic research on health effects is available to all. Thus much of the information that is published in one country can be used beneficially in another. However, each country has its own legislature and its own enforcing authorities and readers should ensure they know their own legislative framework.

Many countries have large organisations concerned with research into health effects, the setting of standards and the dissemination of information. In this section, some major sources of information are listed. While postal addresses are given for most of these, the reader will find that almost all are readily found on the world wide web. Web addresses are given for only a few, due to the problems that arise when addresses change, but the rest may be found very readily using a search engine. The world wide web is an impressive resource. Readers will find that they can now obtain information extremely quickly via the web. However, the material on the world wide web is not peer reviewed and readers should exercise caution. The organisations mentioned in the following sections provide good quality advice, based on sound research, and their publications, and those of organisations like them, are preferred.

10.5.2 International resources

The World Health Organisation (WHO)[37] is concerned with all matters of health and publishes several books of interest to those studying occupational hygiene. The International Agency on Cancer Research (IARC)[38] is part of the WHO and is concerned with the assessment of the data that link substances to cancer. They undertake research of their own and critically assess the evidence available. They maintain a database of all the substances that have been assessed, classified under four headings, according to the weight of evidence. This ranges from those that have been proved to cause cancer, down to those that are probably not carcinogenic. The monographs detailing the evidence that was taken into account are all published on their internet site.

The International Commission on Non-ionising Radiation Protection, ICNIRP,[39] acts as an independent international body of experts whose principal aim is the dissemination of information about the effects of exposure to non-ionising radiations. The International Institute of Welding (IIW)[40] has a Commission (VIII) on the subject of health and safety in welding. They have a limited range of published documents.

10.5.3 The United Kingdom

The text of UK Legislation is obtainable from Her Majesty's Stationery Office (HMSO).[41] The law on health and safety is enforced by the Health and Safety Executive (HSE) for most industrial workplaces and by local authorities for others. The HSE runs an information service and an extensive website.[42] The research arm of the HSE is Health and Safety Laboratory[43] whose research underpins much of the advice offered by the HSE. One of their current projects is to improve the quality of analysis of Cr (VI) in welding fume. This is part of a European proficiency testing scheme previously managed by the Danish External Quality Assessment Scheme.

The enforcement authority for environmental matters is the Environment Agency.[44] Practical advice and guidance on environmental matters are also available from Envirowise.[45]

TWI, The Welding Institute[46] is a non-governmental organisation that carries out research into welding and joining. It offers research, consultancy and advice for its members and offers training facilities to the wider welding community. The website contains many documents giving free advice to the welding community. These consist of a series of sheets with the title 'Job knowledge for welders' and the 'Frequently asked questions' resource. They have also developed an interactive tool 'Welding fume tutor' in conjunction with the Health and Safety Executive, industrial sponsors and union representatives.

The National Radiological Protection Board[31] is concerned with both ionising and non-ionising radiation and gives advice on such diverse subjects as exposure to sunlight, radioactive sources and mobile phones.

The Institution of Occupational Safety and Health[47] is Europe's leading professional body for health and safety professionals. It has a Royal Charter and operates a membership structure designed to reflect the competence levels of its members. It awards the designation Chartered Safety and Health Practitioner CMIOSH, to those who meet the educational and experience requirements.

10.5.4 The USA

The text of legislation is obtainable from the Occupational Safety and Health Administration (OSHA), who enforce the requirements.[48] They develop mandatory safety standards and provide technical assistance, training and education.

The Centers for Disease Control and Prevention (CDC) have the National Institute of Occupational Safety and Health (NIOSH)[49] within its umbrella. This is a federal agency that conducts research and makes recommendations for the prevention of work-related diseases and injury.

The American Conference of Governmental Industrial Hygienists (ACGIH)[50] is a non-governmental organisation of practitioners in industrial hygiene, occupational health, environmental health and safety.

The Board of Certified Safety Professionals[51] is a not-for-profit certification board for safety professionals. It sets the academic and experience standards that are required for practitioners and awards the designation Certified Safety Professional, CSP, to those who meet the educational and experience requirements.

The American Welding Society, AWS,[52] is an organisation that offers certification, research, conferences, education and many other services. It has a publications section that markets a wide range of advice booklets on the subject of welding, a large number of which are concerned with health and safety.

10.5.5 Australia

The text of legislation is published by individual states, but it can be accessed via the Attorney General's Department.[53] The National Occupational Health and Safety Commission is the Statutory authority.[54]

10.5.6 Canada

The Canadian centre for occupational health and safety (CCOHS)[55] is a national organisation giving information about occupational safety and health. It has a resource 'OHS answers' on the world wide web and enables workers in Canada to access the laws specific to their own territory.

10.6 References

1. *Vision for Welding Industry*, 1998, obtainable from American Welding Society, 550 N W LeJeune Rd, Miami FL 33126
2. Blunt J. and Balchin N., *Health and Safety in Welding and Allied Processes*, 5th ed., Cambridge, UK, Woodhead Publishing, 2002
3. *Workplace Exposure Limits*, EH40/2005, available from HSE Books, and from the Internet www.hse.gov.uk/

4. *Occupational Health Statistics Bulletin 2002/3*, available from the statistics section of the HSE website www.hse.gov.uk/statistics
5. Directive 94/9/EC of the European Parliament and the Council, on the approximation of the laws of the Member States concerning equipment and protective systems intended for use in potentially explosive atmospheres, *Official Journal of the European Union*, L100/1, 19 April 1994, 1–29
6. *Dangerous Substances and Explosive Atmospheres Regulations 2002*, Statutory Instrument 2776, 2002, available from HMSO website www.opsi.gov.uk
7. *The Safe Use of Compressed Gases in Welding, Flame Cutting and Allied Processes* HSG 139, Sudbury, UK, HSE Books
8. *Safe Maintenance, Repair and Cleaning Procedures, Approved Code of Practice and Guidance*, L137, Sudbury, UK, HSE Books
9. *Safe Work in Confined Spaces: ACOP Regulations and Guidance*, L101, Sudbury, UK, HSE Books
10. *Odorisation of Bulk Oxygen Supplies in Shipyards* CS7, Sudbury, UK, HSE Books
11. *Safety in Welding, Cutting and Allied Processes*, Z49.1: AWS 1999
12. *Standard for the Safeguarding of Tanks and Containers for Entry, Cleaning and Repair*, NFPA 326. 1999
13. *Criteria for a Standard: Occupational exposure to hand–arm vibration*, DHHS Publication 89-106, September 1989, Washington, DC, NIOSH available from www.cdc.gov/niosh
14. *NIOSH Update: NIOSH pursues hand-vibration studies to understand, address risks*, Internet News release, contact NIOSH Health Effects Laboratory Division
15. *Control of Vibration at Work Regulations 2005*, SI 1093, 2005 available from HMSO website www.opsi.gov.uk/
16. Directive 2002/44/EC of the European Parliament and of the Council, on the minimum health and safety requirements regarding the exposure of workers to the risks arising from physical agents (vibration), *Official Journal of the European Union*, L177 July 2002, 13–20.
17. *Mechanical Vibration. Measurement and evaluation of human exposure to hand-transmitted vibration. General requirements.* BS EN ISO 5349-1 2001. *Mechanical Vibration. Measurement and assessment of human exposure to hand-transmitted vibration. Practical guidance for the measurement at the workplace.* BS EN ISO 5349-2, 2002
18. *Reducing noise at work, Guidance on the Noise at Work Regulations 1989*, L109, Sudbury, UK HSE Books, 1998
19. *Occupational Noise Exposure*, 29 CFR 1910.95
20. *Method of Test for Estimating the Risk of Hearing Handicap due to Noise Exposure.* BS 5330, 1976
21. Directive 2003/10/EC of the European Parliament and of the Council, on the minimum health and safety requirements regarding the exposure of workers to the risks arising from physical agents (noise), *Official Journal of the European Union*, L42, February 2003, 38–44
22. *The Control of Noise at Work Regulation 2005.* Statutory Instrument 1643, 2005, available from HMSO www.opsi.gov.uk/
23. Health effects of welding, *Critical Reviews in Toxicology* **33**(1) 2003, 61–103
24. *NIOSH Strategic Research on Welding Identifies Data Needs, advances studies*, Internet News Release, contact NIOSH Health Effects Laboratory Division
25. Quimby B.J. and Ulrich G.D., Fume formation rates in gas metal arc welding, *Welding Research Supplement*, April 1999, 142s–149s

26. *Fume Emissions from Resistance Welding through Adhesives and Sealants*, Sudbury, UK, TWI Limited, HSE Contract Research Report 388/2001
27. *An Investigation into the Passage of Unburnt Gas through Welding Gaps during the use of Oxy-propane Preheating Torches*, Sudbury, UK, HSE Contract Research Report 78/1995
28. British Compressed Gases Association, 6 St Mary's Street, Wallingford, OX10 0EL. www.bcga.co.uk
29. Compressed Gas Association, 4221 Walney Road, 5th Floor, Chantilly, VA 20151–2923. www.cganet.com
30. *EMF RAPID Questions and Answers – EMF in the workplace*, The National Institute of Environmental Health Sciences September 1996, available from Superintendent of Documents US Government Printing Office Washington, D.C., 20402 (202) 512-1800, and from the Internet
31. National Radiological Protection Board, Chilton, Didcot, Oxon OX11 0RQ, www.nrpb.org.
32. *Amended proposal for a Directive of the European Parliament and of the Council on the minimum health and safety requirements regarding the exposure of workers to the risks arising from physical agents (electromagnetic fields and waves)*, December 2002, available from the Society for Radiological Protection, 76 Portland Place, London W1B 1NT, and from their Internet site www.srp-uk.org
33. Guidelines for limiting exposure to time-varying electric, magnetic and electromagnetic fields up to 300 GHz, *Health Physics* **74**(4) April 1998, 494–522
34. Directive 1999/13/EC Solvent Emissions, on the limitation of emissions of volatile organic solvents in certain activities and installations, *Official Journal of the European Union*, L085, March 1999, 1–22
35. *Vapour Degreasing,* GG015, available from Envirowise, www.envirowise.gov.uk
36. *Our Path to Sustainable Development* London, ESAB, 1999
37. World Health Organisation, www.who.int/en
38. The International Agency on Cancer Research (IARC), www.iarc.fr
39. The International Commission on Non-ionising Radiation Protection, c/o BfS, Ingolstaedter Landstr. 1, 85764 Oberschleissheim, Germany, www.icnirp.de
40. International Institute of Welding, ZI Paris Nord 2, BP: 50362, F95942 ROISSY CDG Cedex, France
41. HMSO address for printed publications: TSO, PO Box 29, St Crispins, Duke Street, Norwich NR3 1GN; full text is also available free of charge from the internet: www.opsi.gov.uk
42. HSE Infoline, Caerphilly Business Park, Caerphilly, CF83 3GG, www.hse.gov.uk
43. Business Development Unit, Health and Safety Laboratory, Broad Lane, Sheffield, S3 7HQ, www.hsl.gov.uk
44. The Environment Agency – see telephone book for the local regional office. www.environment-agency.gov.uk
45. Envirowise, www.envirowise.gov.uk
46. TWI, Granta Park, Great Abington, Cambridge CB1 6AL, www.twi.co.uk
47. Institution of Occupational Safety and Health, The Grange, Highfield Drive, Wigston, Leicester LE18 1NN, www.iosh.co.uk
48. OSHA, US Department of Labor, 200 Constitution Avenue, NW, Washington, DC20210. www.osha.gov
49. NIOSH, 4676 Columbia Parkway, Cincinnati, Ohio 45226, www.cdc.gov/niosh
50. American Conference of Governmental Industrial Hygienists, 1330 Kemper Meadow Drive, Cincinnati, Ohio 45240, USA, www.acgih.org

51. Board of Certified Safety Professionals, 208 Burwash Avenue, Savoy, IL 61874, USA, www.bcsp.org
52. American Welding Society, 550 NW LeJeune Road, Miami, Florida, 33126, USA, www.aws.org
53. Attorney General's Department, Robert Garran Offices, National Circuit, Barton ACT 2600, Australia
54. National Occupational Health and Safety Commission, 92, Parramatta Road, Camperdown, NSW 1460, Australia www.nohsc.gov.au
55. Canadian Centre for Occupational Health and Safety, 250 Main Street East, Hamilton, Ontario L8N 1H6, Canada, www.ccohs.ca

Index

A-TIG 52–64
ablation 67
acid flux 31
activating flux 52–9, 62, 63
active shielding gases 6
adaptive control systems 183, 184
adjustment units 211–12
alkali arc stabilisers 26
aluminium alloy welding
  laser beam 134–8, 174–6, 187–9
  ultrasonic 256–7, 263
aluminium wire electrodes 8–9, 12–13
ammonium nitrate fuel oil (ANFO) 232–3
'anchoring' 246
anode sheaths 44, 45
anodes 200
antireflective coatings 122–3, 125
apertured diaphragm system 225
arc
  constriction 54, 56, 59–60, 186
  currents 186–7
  electromagnetic convection 186
  expansion ratio 47, 73
  instability 35–6
  length 8, 9, 75
  plasmas 54, 55–6, 60, 73
  pressure 48, 51, 60, 65, 67, 73–4
  radius 43
'argon arc' process 40
asperities 254–5
asthma 272–3
ATEX Directive 273
austenite formation 29
austenitic stainless steels 138–40, 171, 173
autogenous welding 95
automation 18, 33, 125, 180, 181, 264–5
axial spray 7, 16
axially directed force 47

'bang and roll' technology 237
barium compounds 30
basic wires 25–6

beam *see* electron beam; laser beam
bimetallic wire 237
binders 25
bipolar cell plates 236
boosters 242, 243, 244, 253
borosilicate crown glass 125
BPP (beam parameter product) 159
brazing 14–15, 167
buoyancy 50, 51

CAD-to-part manufacture 103
carbon contamination 7
carbon dioxide laser welding 81–2, 161–4
  beam delivery 121
  focus spot size 93
  output 85, 86, 159, 160
  shielding gas 94
catenoids 70, 71
cathode stabilisers 32
cathodes 200
cellulose electrodes 37, 38
cementite formation 170
'centre-down' circulation 48, 49
centreline cracking 131, 132
charge-coupled device (CCD) sensors 155
chlorofluorocarbons (CFCs) 283, 284, 285
circulation 48, 49
clad plates 230, 231, 232, 233, 234
clamping systems 134, 144
Class 'I' enclosures 104
coatings 122–3, 125, 130, 133–4, 144, 279
coaxial heads 185
coherence 82
coil joining 99, 101
cold deformation 229
combined welding *see* hybrid welding
component tolerances 76
compressive 'pinch' force 47
conduction
  electrical 41–3
  thermal 44–5, 74, 83, 84, 86–7
conduction mode 84

293

conduction welding 126
constant voltage (CV) power sources 1, 8
contact stresses 258–9
contact tip to work distance (CTWD) 5, 6, 8, 16–17
continuous wave (CW) lasers
  characteristics 116–18
  output 87, 88, 116
  penetration depths 164, 166
convection 83
convective flow 45–6, 58, 59, 60, 61, 62
Converti, J. 46–7, 73
conveyor machines 210
cooling rates 10, 12, 13
cooling systems 103, 119, 125
copper-based alloy welding 143
corrosion resistance 138, 139, 140, 141, 171
costs
  electron beam welding 217, 227
  explosion welding 233, 235
  laser beam welding 100, 102, 103, 120, 133, 191
crack growth rate 13
cracking
  centreline 131, 132
  HAZ 131, 175–6
  hydrogen 30, 38, 170
  random 132
  solidification *see* solidification 'hot' cracking
  transverse 15, 132
  zone 143
'critical frequencies' 248
CTWD (contact tip to work distance) 5, 6, 8, 16–17
CV (constant voltage) power sources 1, 8
CW lasers *see* continuous wave lasers
cycle system machines 208, 209
Czochralski technique 118

deep penetration effect 201–2, 203
'deficit' 49–51
destructive analysis 96–7
detectors 150–4
Diabeam system 220–1, 222, 224, 225
diffraction 88, 89
diffusion-cooled lasers 89, 90
digital control systems 3, 4, 8, 9, 17, 18, 76
dimpled sheets 134
diode laser welding 85, 105–6, 120–1, 167, 168, 177
dip transfer mode 22, 28, 34
disk lasers 168–9
displacement 48–51
dissimilar materials
  electron beam welding 210, 219
  explosion welding 229, 231, 233, 235–6
  laser beam welding 128, 129, 176–7
  ultrasonic welding 257–8, 263–4

double chamber machines 208
doublet lenses 123
droplet transfer 7–8, 25, 29
dual gas GTAW process 67
duplex stainless steels 140–1
DuPont, J. N. 231

eddy currents 3, 155
efflux plasma 66–7
electric and magnetic fields (EMF) 280–2
Electric Power Research Institute 100
electrical conduction 41–3
electrode negative polarity 25, 32
electromagnetic arc oscillation 17
electromechanical conversion 250
electron beam
  constriction 201
  diameter 204, 221, 223–4
  generation 200–1
  manipulation 201, 202, 212–13
  measurement 220–4
  oscillation 205, 213
  quality 220–1
  welding *see* electron beam welding
electron beam welding
  advantages 98–99
  applications 226–7
  keyhole 66–7
  machines 204–10
  micro-electron beam welding 210–14
  non-vacuum (NV-EBW) 214–20
  penetration depth 62
  process 200–6
  quality assurance 220–6
electron bombardment 200
electron emission 201, 203
electronic packaging 237
electroslag welding 36–7
Elenbaas-Heller equation 43
energy absorption efficiency 85–6
energy transport in GTAW 41–6, 74
environmental issues 76–7, 191, 282–6
'essential variables' 94
excited droplet oscillation 9
explosion bulge test 97
explosion hazards 273, 279–80
explosion welding (EXW)
  applications 233–8
  bond morphology 238–9
  capabilities 229–30
  processes 231–3
exposure limits 271–2, 276, 277, 282

'fall' voltages 44, 45
'far field' welding 268
fast beam welding 205–6
fatigue life 11, 12, 13
FBTIG (flux bounded TIG) 63–4
feed-forward controls 8

Index 295

ferritic stainless steels 140, 173–4
fiber lasers
   applications 104–5
   beam quality 105
   characteristics 122, 159, 160, 161
   penetration depth 164, 165
   pumping systems 169, 170
fiber optics 87, 93, 99, 123, 124
field emission 44
filler metal 95, 107, 135–6, 185–6, 265
fillet welding 27, 30, 33
flux *see* activating flux
flux bounded TIG (FBTIG) 63–4
flux-cored wires 21, 22–3, 24, 25, 30, 33, 35
flyer plates 230
focal spot size 91–2, 99, 105, 122, 123, 124–5
focus position 145
focusing lenses 82, 91, 92, 122, 124, 201
'free surfaces' 51, 68–72
frequency tracking circuitry 245, 248–9
friction reduction 3, 5
fume hazards 271–3, 278–9

galvanometric scanners 106
gap bridgeability 14, 128, 129, 130, 171
gas
   flow 94–5, 127
   mixes 6–7
   shielding *see* shielding gases
   trails 36
   viscosity 48, 74
'gasless electrogas' welding 29
Gaussian distribution 48, 88, 89, 90
GGTAW (guided GTAW) 77
Glickstein, S.S. 43
global warming 284–5
GTAW (gas tungsten arc welding)
   keyhole 64–76
   principles 41–52
   process 52–64

HAZ cracking 131, 175–6
health and safety 30, 76–7, 103–4, 216, 232–3
   environmental issues 282–6
   legislation 271–7, 286–8
   research 277–82
heat radiation 3, 5
heat transfer 61, 62
heavy plate fabrication 16–17, 29, 226
'heliarc' process 40
hermetic seals 115, 133, 142
hermiticity checking 97
high current GTAW 64–5
high-energy density 37, 77
'high spots' *see* asperities
'hot' arcs 44, 45
hot cracking *see* solidification cracking
'hot wire' feed 107
'humping' instability 107

hybrid welding 13–17, 77, 90, 101, 105, 185–9
hydrogen absorption 36
hydrogen cracking 30, 38, 170
hygroscopic synthetic titanates 27
hysteresis 72

in-process monitoring *see* monitoring techniques
in-process repair 183, 184
in-situ observations 97–8, 171
infrared black body emission 149
internet 17
interstitial embrittlement 141
inverse bremsstrahlung absorption 86
'inverter' technology 3, 4, 23

joint configurations 128, 129, 145, 146, 211
joint tracking systems 153–5
Joule resistance 200

keyhole welding
   GTAW 64–76
   laser 66–7, 84, 85–6, 126–7, 178–80
Kjellberg, Oscar 21
Kyoto Climate Change Protocol 285

LAN (local area networks) 17
Laplace's equation 67–8
laser beam
   hazards 103, 104
   output 82–3, 128
   positioning 95–6
   quality 90–1, 92, 159–60, 161, 162, 180–1
   scanning 106
   'waist' 89, 90, 159, 161
   welding *see* laser beam welding
laser beam welding (LBW)
   advances 180–9
   advantages 98–9, 126, 158
   applications 100–2, 189–90
   characteristics 160
   energy efficiency 85–7
   keyhole 66–7, 84, 85–6, 126–7, 178–80
   machining systems 121–5
   Nd:YAG *see* Nd:YAG laser welding
   parameters 87–96
   process 14, 15, 37, 129–32
   quality assurance 96–8, 144–55
   research 170–80, 191
   safety 103–4
laser blazing 189
laser cladding 102
laser hardfacing 102
Laser Institute of America 108
Laser Safety Officers 104
laser weld repair 102, 103
lateral drive systems 242, 252
lime-fluorspar systems 25, 26, 28–9

lock chamber machines 208
longitudinal modes 88
Lorentz force 46, 47, 48, 60

MAG welding 6
magnetic fields 280–2
magnification factor 'M' 89, 123
Maiman, Thomas 81, 158
manual metal arc (MMA) electrodes 21, 24, 25, 33
marangoni flow 48, 58, 60–2, 83, 186
martensite formation 37, 170
martensitic stainless steels 140
MC (metal core) wires 5, 6, 21, 27–8, 33
melting efficiency 86–7
MEMS (micro-electro-mechanical systems) 18
metal core (MC) wires 5, 6, 21, 27–8, 33
metal pairs, welding *see* dissimilar materials
metallurgical control 25, 29, 95, 105
methyl ethyl ketone (MEK) 53
Metzbower, E. A. 98
micro-electro-mechanical systems (MEMS) 18
micro-electron beam welding 210–14
micro-fusion 232
microalloying 27
microstructure-hardening 170
microwelds 255, 256
MIG welding 6
MMA (manual metal arc) electrodes 21, 24, 25, 33
momentum transport 46–8
monitoring techniques 148–52, 180–3
   see also in-situ observations
multi axis motion system 87
multipass welding 31, 40
multiple beam welding 107–8, 134, 205–6, 213

narrow groove welding 17, 63
National Institute for Occupational Safety and Health (NIOSH) 274, 275, 277
Navier-Stokes equation 46
Nd:glass lasers 81
Nd:YAG (Neodynium doped Yttrium Aluminium garnet) laser welding
   applications 99, 105
   control 144–55
   design 118–20
   energy efficiency 85, 86
   machining 121–5
   materials 132–43
   output 82, 92, 93, 113–21
   penetration depths 164–6
   process 125–9
   process development 129–32
   safety 104
   shielding gas 94
nickel-based alloy welding 143
noise damage 276–7
non-destructive examination (NDE) 97

non-vacuum electron beam welding (NV-EBW) 214–20, 217–19

ODPP (one-drop-per-pulse) method 9
optic systems 204
optical sensors 149
orbital welding 63
oxygen content in flux 58, 60
oxygen-low weld metals 25, 26, 31
ozone 282–4

pacemakers 141–2, 281
Paris law 13
Patel, C. K. N. 81
peak power (PP) 113–17, 250
'pilot arc' 77
'pinch' force 47
pinholes 132
pipework *see* tube welding
planar welds 229, 230
plasma arc welding (PAW) 40, 52, 66, 67
plasma formation 162, 163, 171, 216
plasma jets 46, 47, 67
plasma sensors 155
plasma suppression jets 86, 94
plastic deformation 255, 256, 257, 263
plastics, welding 177, 178, 214, 241, 247, 268
porosity
   aluminium alloy welding 137, 174–5
   carbon dioxide laser welding 163–4
   GTAW 72
   hybrid welding 187, 188
   laser welding 177–80
   metal arc welding 11, 12
   pulsed laser welding 164, 166
   tubular cored wire welding 25, 28, 36
power densities
   electron beam welding 204, 221
   GTAW 41, 67
   laser beam welding 87–8, 113, 114, 123
   tubular cored wire welding 37
power sources 1–3, 23, 245
prefabricated primers 24
preheating 106–7, 132, 279
pressing quality 144
professional bodies 108, 287–8
pulse energy 114–5
pulse frequency 115, 116
pulsed current welding 3–4, 9–13, 23
pulsed lasers
   applications 99, 102, 190
   output 82, 113, 114–8
   parameters 87–8, 92–3
   penetration depths 164, 166
pulsed spray 7, 16
pump cavity configurations 118–19
punch-and-die forming 101
push-pull system 3, 5, 6, 7

Q-mass analysis 175

radial pressure gradient 47
radiation 83, 104, 216, 227
radio frequency (RF) generator 81
random cracking 132
reactive thermal conductivity 44–5, 74
recirculatory flow 60, 62
recoil pressures 67
reflectivity 85, 87, 135, 160, 174
regression analysis 9
relative motion 241, 242, 246
remote laser welding 180, 181, 189
repetition rate 92, 99, 102, 116
research groups 108–9
resonant frequencies 245, 248
resonators 88–90
retro-reflecting mirrors 88, 89, 90
reverse machining 102
ring welds *see* torsion welds
robotic systems *see* automation
rollbonding 233, 235
rotating wire sensors 221, 225–6
rutile flux-cored wires 23, 24, 26–7, 33, 35

safety *see* health and safety
scanner laser welding 180, 181, 189
scanning electron microscopy (SEM) 204, 210–2, 238
'scraper' mirrors 89
seam tracking 15, 148, 155
seam welding 252–3, 262
seamless tubular wires 32–3
secondary hardening 29
Seebeck calorimeters 86
self-shielded wires 28–30, 35–7
SEM (scanning electron microscopy) 204, 210, 211–2, 238
semi-automatic welding 29, 37
sensors
    charge-coupled device (CCD) 155
    optical 149
    plasma 155
    rotating wire 221, 225–6
    slit-hole 220–1, 224–5
    tactile 153
    vision based 155
shielding gases
    choosing 28, 94, 127, 162, 171–3
    functions 41, 145–7, 175
    properties 43
    types 6–7, 141, 147–8
shearing 99, 241, 242, 247, 254, 255, 259
sheath regions 43–4
signal acquisition 18
single-knob ('synergic') adjustment 3, 4
slit-hole sensors 220–1, 224–5
solidification 'hot' cracking
    aluminium alloy welding 135–6, 137, 138, 175–6, 190
    stainless steel welding 170, 173–4
solidification mechanisms 10

solidification rates 103
Solvent Emissions Directive 284
sonotrodes 242–4, 252
space charge 44
spatter reduction 23, 128, 164
spherical aberration 125
spot welding 117, 118, 178–9, 183, 184, 238
spray transfer 22
sputtering targets 237
stable mode 88–9
stable-unstable mode 90
stainless steel cladding 13, 14
stainless steel welding 138, 143, 173–4, 189, 236
standards, welding 109–110
static clamping force 247, 249, 254, 259
stationary beams 123
steel welding 133–4, 170–1, 236
    see also stainless steel welding
'step index' fibres 122
stereolithography 103
stick electrodes 24, 26
stimulated emission 82
strip wires 6, 7
structural steel welding 143
subharmonic vibrations 266
submerged arc welding 31
sulphur content in flux 58, 59, 60, 62
supermodulation (SM) 116, 117, 131
surface active elements 58–9, 60, 61, 62
surface curvature 67–71
surface distortion 51
surface quality 144, 264
surface temperature 56
surface tension
    A-Tig 57–8, 60, 61, 62
    GTAW 48, 51
    keyhole welding 67–73
    laser beam welding 83–4
surface tension droplet transfer 7, 8

tactile sensors 153
tailor blank welding
    applications 97, 99, 100–1, 217
    process 133, 149–50, 170
tandem welding 15–17, 28, 33, 37
'tears of wine' phenomena 62
TEM (transmission electron microscopy) 238
tensile strength 10, 12, 13, 137, 176
terminating craters 49, 66
terminating electrodes 47
thermal conduction 44–5, 74, 83, 84, 86–7
thermal deterioration 3, 5, 6
thermal distortion 93, 125, 138, 141
thermal ionisation 41
'thermal pinch' 67
thermionic emission 44, 200
threshold current 73–5
through-arc sensing 28
through-hole defects 182–4

titanium alloy welding 141–2
titanium nitride formation 29
torsion welding 252–4, 262, 267
total curvature 68
total internal reflection 122
transducers 242–5, 248, 252–4
transformers 3, 23
transition joints (TJ) 235–6
transmission electron microscopy (TEM) 238
transmission loss 122
transverse cracking 15, 132
transverse modes 88
transverse shearing force 247
transverse vibrations 246–7, 255
triode systems 200
tube welding 63, 141, 230, 231
tubular electrodes
    advantages 5–6, 24–5, 36–8
    applications 34–5
    disadvantages 35–8
    equipment *22*, 23
    manufacturing 31–3, *32*
    materials 25–33
    process control 34
tungsten welding *see* GTAW
tunnel porosity 72
twin beam welding 107–8, 134

ultrasonic frequency 242, 243, 248–9
ultrasonic testing 34, 97, 155
ultrasonic welding
    advantages 262–6

applications 260–2
equipment 252–4
mechanics 254–60
principles 242–52
underwater welding 189, 190
unstable modes 89
upward melt flow 180

vacuum chambers 204, 206, 207–10, 233
vapour cavities 201–3
vapour concentration 56
venting 67
vibration amplitude 243–5, 249
vibration damage 274–6
vision based sensors 155
visual examinations 96

waveguide mode 89
wedge-reed system 245–6, 252
Wehnelt cylinders 200
weld overlay 233, 235
weld pool behaviour 48–52, 57
wire electrode geometry 5–6
wire feeding 3, 5, *5*, 23, 35–6
'worm trails' 36

X-rays 97, 171
X-ray radiation 216, 227

Zinc-coated steel welding 171–3, 189
zone cracking 143
zoning 273